WORKBOOKS IN CHEMISTRY

INORGANIC CHEMISTRY

Matthew Almond
University of Reading

Mark Spillman
University of Reading

Elizabeth Page
University of Reading

Series editor:
Elizabeth Page
University of Reading

OXFORD
UNIVERSITY PRESS

OXFORD
UNIVERSITY PRESS

Great Clarendon Street, Oxford, OX2 6DP,
United Kingdom

Oxford University Press is a department of the University of Oxford.
It furthers the University's objective of excellence in research, scholarship, and education by publishing worldwide. Oxford is a registered trade mark of Oxford University Press in the UK and in certain other countries

Impression: 1

Published in the United States of America by Oxford University Press
198 Madison Avenue, New York, NY 10016, United States of America

British Library Cataloguing in Publication Data
Data available

ISBN 978-0-19-872950-1

Printed in Great Britain by
Bell & Bain Ltd., Glasgow

Periodic table of the elements

s-block · d-block · p-block

Period	Group 1	2	3	4	5	6	7	8	9	10	11	12	13	14	15	16	17	18
1							1 **H** 1.0079											2 **He** 4.0026
2	3 **Li** 6.941	4 **Be** 9.0122											5 **B** 10.811	6 **C** 12.011	7 **N** 14.007	8 **O** 15.999	9 **F** 18.998	10 **Ne** 20.180
3	11 **Na** 22.990	12 **Mg** 24.305											13 **Al** 26.982	14 **Si** 28.086	15 **P** 30.974	16 **S** 32.065	17 **Cl** 35.453	18 **Ar** 39.948
4	19 **K** 39.098	20 **Ca** 40.078	21 **Sc** 44.956	22 **Ti** 47.867	23 **V** 50.942	24 **Cr** 51.996	25 **Mn** 54.938	26 **Fe** 55.845	27 **Co** 58.933	28 **Ni** 58.693	29 **Cu** 63.546	30 **Zn** 65.409	31 **Ga** 69.723	32 **Ge** 72.64	33 **As** 74.922	34 **Se** 78.96	35 **Br** 79.904	36 **Kr** 83.798
5	37 **Rb** 85.468	38 **Sr** 87.62	39 **Y** 88.906	40 **Zr** 91.224	41 **Nb** 92.906	42 **Mo** 95.94	43 **Tc** (98)	44 **Ru** 101.07	45 **Rh** 102.91	46 **Pd** 106.42	47 **Ag** 107.87	48 **Cd** 112.41	49 **In** 114.82	50 **Sn** 118.71	51 **Sb** 121.76	52 **Te** 127.60	53 **I** 126.90	54 **Xe** 131.29
6	55 **Cs** 132.91	56 **Ba** 137.33	57 **La** 138.91	72 **Hf** 178.49	73 **Ta** 180.95	74 **W** 183.84	75 **Re** 186.21	76 **Os** 190.23	77 **Ir** 192.22	78 **Pt** 195.08	79 **Au** 196.97	80 **Hg** 200.59	81 **Tl** 204.38	82 **Pb** 207.2	83 **Bi** 208.98	84 **Po** (209)	85 **At** (210)	86 **Rn** (222)
7	87 **Fr** (223)	88 **Ra** (226)	89 **Ac** (227)	104 **Rf** (263)	105 **Db** (262)	106 **Sg** (266)	107 **Bh** (272)	108 **Hs** (277)	109 **Mt** (276)	110 **Ds** (281)	111 **Rg** (280)	112 **Cn** (277)	113 **Nh** unknown	114 **Fl** (289)	115 **Mc** unknown	116 **Lv** (298)	117 **Ts** unknown	118 **Og** unknown

f-block

Lanthanides 6	58 **Ce** 140.12	59 **Pr** 140.91	60 **Nd** 144.24	61 **Pm** (145)	62 **Sm** 150.36	63 **Eu** 151.96	64 **Gd** 157.25	65 **Tb** 158.93	66 **Dy** 162.50	67 **Ho** 164.93	68 **Er** 167.26	69 **Tm** 168.93	70 **Yb** 173.04	71 **Lu** 174.97
Actinides 7	90 **Th** 232.04	91 **Pa** 231.04	92 **U** 238.03	93 **Np** (237)	94 **Pu** (244)	95 **Am** (243)	96 **Cm** (247)	97 **Bk** (247)	98 **Cf** (251)	99 **Es** (252)	100 **Fm** (257)	101 **Md** (258)	102 **No** (259)	103 **Lr** (262)

Key

8 — Atomic number
O — Symbol
15.999 — Relative atomic mass

Tables of data

Physical constants

Name	Symbol	Value
Avogadro constant	N_A	6.022×10^{23} mol^{-1}
Ideal gas constant	R	8.314 J K^{-1} mol^{-1}
Boltzmann constant	k_B	1.381×10^{-23} J K^{-1}
Planck constant	h	6.626×10^{-34} J s
Faraday constant	F	96 485 C mol^{-1}
Rydberg constant	R_H	1.097×10^{7} m^{-1}
Kapustinskii constant	κ	1.0790×10^{-4} J m mol^{-1} 107 900 kJ pm mol^{-1}
Speed of light in vacuum	c	2.998×10^{8} m s^{-1}
Elementary charge	e	1.602×10^{-19} C
Electron mass	m_e	9.109×10^{-31} kg
Proton mass	m_p	1.673×10^{-27} kg
Neutron mass	m_n	1.675×10^{-27} kg
Permittivity of free space	ε_0	8.854×10^{-12} J^{-1} C^{2} m^{-1}

List of SI prefixes

Name	Symbol	Value
Pico	p	$\times 10^{-12}$
Nano	n	$\times 10^{-9}$
Micro	μ	$\times 10^{-6}$
Milli	m	$\times 10^{-3}$
Kilo	k	$\times 10^{3}$
Mega	M	$\times 10^{6}$
Giga	G	$\times 10^{9}$
Tera	T	$\times 10^{12}$

Overview of contents

Preface

Welcome to the Workbooks in Chemistry

The Workbooks in Chemistry have been designed to offer additional support to help you make the transition from school to university-level chemistry. They will also be useful if you are studying for related degrees, such as biochemistry, food science, or pharmacy.

Introduction to the Workbooks

The Workbooks cover the three traditional areas of chemistry: inorganic, organic, and physical. They are designed to complement your first year chemistry modules and to supplement, but not replace, your course text book and lecture notes. You may want to use the Workbooks as self-test guides as you carry out a specific topic, or you may find them useful when you have finished a topic as you prepare for end of semester tests and exams. When preparing for tests and exams, students often use practice questions, but model answers are not always available. This is because there is usually more than one correct way to answer a question and your lecturers will want to give you credit for your problem-solving approach and working, as well as having obtained the correct answer. These Workbooks will give you guidance on good practice and a logical approach to problem solving, with plenty of hints and tips on how to avoid typical pitfalls.

Structure of the Workbooks

Each of the three Workbooks is divided into chapters covering the different topics that appear in typical first year chemistry courses. As external examiners and assessors at different UK and international universities, we realize that every chemistry programme is slightly different, so you may find that some topics are covered in more depth than you require, or that there are topics missing from your particular course. If this is the case, we would be interested in hearing your views! However, we are confident that the topics covered are representative, and that most first year students will meet them at some point.

Each chapter is divided into sections, and each section starts with a brief introduction to the theory behind the concepts to put the subsequent problems in context. If you need to, you should refer to your lecture notes and text books at this point to fully revise the theory.

Following the outline introduction to each topic, there is a series of **worked examples**, which are typical of the problems you might be asked to solve in workshops or exams. These examples contain fully worked solutions that are designed to give you the scaffolding upon which to base any future answers, and sometimes provide you with hints about how to approach these types of question and how to avoid common errors.

After the worked examples relating to a topic, you will find further **questions** of a similar type for you to practise. The numerical or 'short' answers to these problems can be found at the end of the book, whilst fully worked solutions with extensive explanations in most cases are available on the Online Resource Centre. At the end of each book is a bank of **synoptic questions**, also with worked solutions on the Online Resource Centre. Synoptic questions encourage you to draw on concepts from multiple topics, helping you to use your broader chemical knowledge to solve problems.

You can find the series website at http://www.oxfordtextbooks.co.uk/orc/chemworkbooks.

How to use the Workbooks

You will probably refer to these Workbooks at different times during your first year course, but we envisage they will be most useful when preparing for examinations after you have done some initial revision.

It is a good idea to use the introductions to the topics to check your understanding and refresh your memory. The next step is to follow through the worked examples, or try them out yourself.

The **hints** will give you guidance on how to tackle the problem—for example, reminding you of points you may need to use from different areas of chemistry.

Question 2.3

Would you expect to see more, or fewer, examples of σ–π crossover for the homonuclear diatomic molecules of the third row (Na_2–Cl_2) of the periodic table, than we see with the homonuclear diatomic molecules of the second row (Li_2–F_2)?

► **Hint** In your answer, consider how the difference in energy between adjacent orbitals changes as we go down the periodic table.

The **comments** will typically relate to the worked solutions and might explain why a unit conversion has been used, for example, or give some background explanation for the maths used in the solution. The comments are designed to help you avoid the typical mistakes students make when approaching each particular type of problem. It is to be hoped that by being aware of these pitfalls you will be able to overcome them.

In this structure the S^{2-} ions are in the same positions as the Na^+ ions in NaCl (or the Cl^- ions) but in zinc blende the counter ions occupy half of the tetrahedral holes in the unit cell.

You can check that the ratio is correct as the charge on Zn = +2 and the charge on S = –2 so the charge is balanced.

When you are happy you have mastered the worked examples, try the questions. To check your answers go to the back of the book, and to check your working look for the fully worked solutions at http://www.oxfordtextbooks.co.uk/orc/chemworkbooks.

The synoptic questions can be used as a final revision tool when you are confident with your understanding of the individual topics and want some final practice before the exam or test. Again, you will find answers at the back of the book and full solutions online.

Final comments

We hope you find these Workbooks helpful in reinforcing your understanding of key concepts in chemistry and providing tips and techniques that will stay with you for the rest of your chemistry degree course. If you have any feedback on the Workbooks—such as aspects you found particularly helpful or areas you felt were missing—please get in touch with us via the Online Resource Centre. Go to http://www.oxfordtextbooks.co.uk/orc/chemworkbooks.

1

Atomic structure

1.1 Electromagnetic waves and matter

Chemists frequently make use of **spectroscopic** techniques to study materials. Such techniques rely on the interaction between electromagnetic waves and matter. Electromagnetic waves travel in a vacuum at the **speed of light**; a constant speed of 2.998×10^8 m s^{-1}. This universal constant is given the symbol c. As well as its speed, an electromagnetic wave is also described by its **wavelength** and its **frequency**. The wavelength is the distance between consecutive crests or troughs in the wave and has units of metres (m). It is represented by the Greek letter λ (lambda), while the frequency is the number of wave crests or troughs that pass over a fixed point in one second and so has units of Hertz; Hz or s^{-1}. It is represented by the Greek letter ν (nu). These quantities are related as follows:

$$\nu = \frac{c}{\lambda}$$

For mostly historical reasons, spectroscopists also frequently make use of another related quantity known as the **wavenumber**. It is denoted by the symbol $\bar{\nu}$ (pronounced 'nu-bar'), and is defined as:

$$\bar{\nu} = \frac{1}{\lambda}$$

which gives the number of wavelengths that add together to give one metre.

In classical physics, electromagnetic radiation is thought of and treated mathematically as a wave. In 1900 a revolutionary idea was introduced by Max Planck, who solved a long-standing problem in classical physics and sparked the development of Quantum Theory (see black body radiation, Burrows et al., page 120). His solution to this problem involved the proposal that electromagnetic radiation can only be emitted in discrete packets or **quanta**. These packets of light later became known as **photons**, the energy of which can be calculated from their frequency, according to the Planck relation:

$$E = h\nu$$

where h is the Planck constant (6.626×10^{-34} J s) and ν is the frequency of the photon. This suggested that light can behave both as a particle and as a wave.

In the mid-1920s, Louis de Broglie formulated the hypothesis that **all matter** has both particle- and wave-like characteristics. He proposed an equation which has since been rigorously experimentally verified, that relates the wavelength of a particle or object to its momentum:

$$\lambda = \frac{h}{p}$$

where h is the Planck constant and p is the momentum of the particle. This remarkable result forms the basis for a number of analytical techniques in use today.

Worked example 1.1A

Photons emitted in a spectroscopic transition have a wavelength of 450 nm. Calculate the following quantities, and give the correct units for your answers:

(a) The wavenumber of these photons.

(b) The frequency of these photons.

(c) The energy of **one** of these photons.

(d) The energy of **one mole** of these photons, giving your answer in kJ mol^{-1}.

► **Hint** A good habit to develop when performing calculations in chemistry is to convert all values given into SI units and then, if necessary, convert back again after completing the numeric manipulations.

Solution

The photon wavelength given is 450 nm. One nanometre is equal to 1×10^{-9} m, so the wavelength given is equivalent to 450×10^{-9} m which in standard form is 4.50×10^{-7} m.

(a)

$$\bar{\nu} = \frac{1}{\lambda}$$

$$= \frac{1}{4.50 \times 10^{-7}\,\mathrm{m}}$$

$$= 2.22 \times 10^{6}\,\mathrm{m}^{-1}$$

Remember that $\bar{\nu} = \frac{1}{\lambda}$ and $\nu = \frac{c}{\lambda}$.

(b)

$$\nu = c\bar{\nu}$$

$$= 2.998 \times 10^{8}\,\mathrm{m\,s^{-1}} \times 2.22 \times 10^{6}\,\mathrm{m}^{-1}$$

$$= 6.66 \times 10^{14}\,\mathrm{s}^{-1}$$

(c)

$$E_{\mathrm{photon}} = h\nu$$

$$= 6.626 \times 10^{-34}\,\mathrm{J\,s} \times 6.66 \times 10^{14}\,\mathrm{s}^{-1}$$

$$= 4.41 \times 10^{-19}\,\mathrm{J}$$

(d)

$$E_{\mathrm{mole}} = E_{\mathrm{photon}} \times N_{\mathrm{A}}$$

$$= 4.41 \times 10^{-19}\,\mathrm{J} \times 6.022 \times 10^{23}\,\mathrm{mol}^{-1}$$

$$= 2.66 \times 10^{5}\,\mathrm{J\,mol}^{-1}$$

$$= 266\ \mathrm{kJ\,mol}^{-1}$$

Worked example 1.1B

Calculate the de Broglie wavelengths of a proton and an electron travelling at one-tenth of the speed of light. Give your answers in picometres.

Solution

First, we calculate the velocity, v, at which the particles are travelling.

$$v = 0.1c$$
$$= 0.1 \times 2.998 \times 10^{8}\,\text{m s}^{-1}$$
$$= 2.998 \times 10^{7}\,\text{m s}^{-1}$$

We can now calculate the momentum possessed by each particle by using this velocity and the masses given in the list of physical constants (see Tables of data at the front of the book). These values can then be used in the de Broglie equation to calculate the particle wavelengths.

$$p_{\text{proton}} = 1.673 \times 10^{-27}\,\text{kg} \times 2.998 \times 10^{7}\,\text{m s}^{-1}$$
$$= 5.016 \times 10^{-20}\,\text{kg m s}^{-1}$$

$$p_{\text{electron}} = 9.109 \times 10^{-31}\,\text{kg} \times 2.998 \times 10^{7}\,\text{m s}^{-1}$$
$$= 2.731 \times 10^{-23}\,\text{kg m s}^{-1}$$

We can now calculate the wavelengths of the proton and electron using the de Broglie equation:

$$\lambda = \frac{h}{p}$$

$$\lambda_{\text{proton}} = \frac{6.626 \times 10^{-34}\,\text{J s}}{5.016 \times 10^{-20}\,\text{kg m s}^{-1}} = \frac{6.626 \times 10^{-34}\,\text{kg m}^2\,\text{s}^{-2}\,\text{s}}{5.016 \times 10^{-20}\,\text{kg m s}^{-1}} = 1.321 \times 10^{-14}\,\text{m}$$

$$\lambda_{\text{electron}} = \frac{6.626 \times 10^{-34}\,\text{J s}}{2.731 \times 10^{-23}\,\text{kg m s}^{-1}} = \frac{6.626 \times 10^{-34}\,\text{kg m}^2\,\text{s}^{-2}\,\text{s}}{2.731 \times 10^{-23}\,\text{kg m s}^{-1}} = 2.426 \times 10^{-11}\,\text{m}$$

Now we convert our answers to picometres using: $1\,\text{pm} = 1 \times 10^{-12}\,\text{m}$

$$\lambda_{\text{proton}} = 1.321 \times 10^{-2}\,\text{pm}$$

$$\lambda_{\text{electron}} = 24.26\,\text{pm}$$

> The similarity between the Greek letter **nu** (ν) used for frequency, and the letter v used for velocity can be confusing. Make sure you are aware of when these symbols should be used, and take care when writing them down to ensure your answers are legible and not misleading.

> Momentum, p, is mass multiplied by velocity: $p = mv$

> The joule is a **derived unit** of energy, work, or heat in the SI unit system. It is defined formally as $1\,\text{J} = 1\,\text{kg m}^2\,\text{s}^{-2}$. It can be seen from this expanded form that the units of the Planck constant and momentum cancel out by division to give metres—the expected unit for wavelength. The use of units allows us to check that our working makes sense.

Question 1.1

One of the most important processes that occurs in the Earth's upper atmosphere is the photodissociation of oxygen molecules into atomic oxygen:

$$O_2\,(g) + h\nu \rightarrow 2O\,(g)$$

The minimum energy a photon can have for this process to occur is equivalent to the bond dissociation enthalpy in the oxygen molecule. Given that the bond enthalpy of O_2 is 495 kJ mol^{-1}, calculate the wavelength of a photon that has the minimum energy required to break this bond, giving your answer in nanometres.

> The bond enthalpy is a measure of the strength of a bond, and gives us the energy required to break a particular covalent bond. It is normally given as a molar quantity.

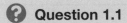

Question 1.2

Diffraction is a phenomenon that occurs when a wave meets an object, or gap in an object. The technique of electron diffraction exploits the wave-like properties of electrons, and observes the resultant **interference** pattern that arises when electrons pass through crystalline solids or gaseous molecules.

The effect is strongest when the wavelength is roughly equivalent to the size of the diffracting object (or gap). The average distance between neighbouring ions in a crystal of table salt (NaCl) is 282 pm. By assuming that the optimal wavelength of the electrons is equal to this distance, calculate the ideal speed for the electrons in this experiment.

Would the equivalent neutron diffraction experiment require higher or lower speeds?

Note: the mass of a neutron is 1.675×10^{-27} kg.

1.2 The photoelectric effect

Albert Einstein is best known for his theories of special and general relativity. However, his Nobel Prize was awarded for his explanation of the photoelectric effect. This effect is observed when electromagnetic radiation of a sufficiently high energy strikes a metal surface. Different types of metals have different threshold limits on the energy required, but in each case, no matter how intense the light, if the photons do not have energies above the threshold, no electrons will be ejected.

Einstein reasoned that electrons can only be ejected from the surface when they have absorbed sufficient energy from the incoming light. Once above this threshold, the excess energy of the photon that was transferred to the electron is converted into kinetic energy. He showed that the energy of the incident photons could be calculated from their frequency and wavelength using the Planck relation seen in section 1.1, and therefore, the kinetic energy of the electrons released could be calculated using:

$$E_{\mathrm{KE}} = hv - \Phi$$

where E_{KE} is the kinetic energy of the ejected electron, h is the Planck constant, v is the photon frequency, and Φ is the **work function** of the material (i.e. the threshold energy required to remove electrons from a given material). This is the Greek letter phi.

Worked example 1.2A

Lithium metal has a work function of 283 kJ mol^{-1}. Calculate:

(a) the speed (in km s^{-1}) and (b) the de Broglie wavelength (in nm) of electrons ejected by 300 nm ultra-violet light.

Solution

We first calculate how much energy is required to liberate a single electron:

$$\Phi_{\mathrm{electron}} = \frac{\Phi_{\mathrm{mole}}}{N_{\mathrm{A}}} = \frac{283 \times 10^{3}\,\mathrm{J\,mol^{-1}}}{6.022 \times 10^{23}\,\mathrm{mol^{-1}}} = 4.70 \times 10^{-19}\,\mathrm{J}$$

Next, we calculate the energy of a photon with $\lambda = 300$ nm:

$$E_{\mathrm{photon}} = hv = \frac{hc}{\lambda} = \frac{6.626 \times 10^{-34}\,\mathrm{J\,s} \times 2.998 \times 10^{8}\,\mathrm{m\,s^{-1}}}{300 \times 10^{-9}\,\mathrm{m}} = 6.62 \times 10^{-19}\,\mathrm{J}$$

We now have the necessary information to calculate the kinetic energy of an ejected electron using the following equation:

$$E_{KE} = hv - \Phi = E_{photon} - \Phi = 6.62\times10^{-19}\,J - 4.70\times10^{-19}\,J = 1.92\times10^{-19}\,J$$

Finally, remembering that kinetic energy is given by $E_{KE} = \frac{1}{2}mv^2$ and momentum by $p = mv$, we can calculate the velocity and momentum of the electron. This can then be used to calculate the de Broglie wavelength using $\lambda = \frac{h}{p}$.

(a) The velocity of the electron may be obtained by rearranging the equation for kinetic energy to make v the subject.

$$v = \sqrt{\frac{2E_{KE}}{m_e}}$$

$$= \sqrt{\frac{2\times1.92\times10^{-19}\,J}{9.109\times10^{-31}\,kg}}$$

$$= \sqrt{\frac{2\times1.92\times10^{-19}\,kg\,m^2\,s^{-2}}{9.109\times10^{-31}\,kg}}$$

$$= \sqrt{4.22\times10^{11}\ m^2\,s^{-2}}$$

$$= 6.49\times10^{5}\ m\,s^{-1}$$

$$= 649\ km\,s^{-1}$$

(b) The momentum of such an electron can then be obtained by multiplying the velocity by m_e (the mass of an electron).

$$p_{electron} = m_e v$$

$$= 6.49\times10^{5}\ m\,s^{-1}\times9.109\times10^{-31}\ kg$$

$$= 5.91\times10^{-25}\ kg\,m\,s^{-1}$$

Now we can calculate the wavelength of such an electron using the de Broglie relation, and then convert our answer to nanometres to satisfy the question:

$$\lambda = \frac{h}{p}$$

$$= \frac{6.626\times10^{-34}\,J\,s}{5.91\times10^{-25}\,kg\,m\,s^{-1}}$$

$$= \frac{6.626\times10^{-34}\,kg\,m^2\ s^{-2}\ s}{5.91\times10^{-25}\,kg\,m\,s^{-1}}$$

$$= 1.12\times10^{-9}\ m$$

$$= 1.12\ nm$$

We can check our working using units by remembering that the joule is a **derived unit** of energy, work, or heat in the SI unit system. It is defined formally as: 1 J = 1 kg m^2 s^{-2}.

Question 1.3

Photons with a wavelength of 169 nm are able to eject electrons from mercury metal. These electrons have a wavelength of 727 pm. Using this information, calculate the work function of mercury, giving your answer in kJ mol^{-1}.

Question 1.4

Caesium metal has the lowest work function of any of the group 1 metals: 202.6 kJ mol^{-1}.

(a) Calculate the wavelength of light required to eject electrons from caesium that have a kinetic energy of 30 kJ mol^{-1}. To which part of the electromagnetic spectrum do these photons belong?

(b) At what percentage of the speed of light are these electrons travelling?

1.3 Hydrogenic emission spectra and the Rydberg equation

Remember that the energy of a photon, E, is proportional to its frequency, ν, according to the Planck relation $E = h\nu$. This can also be used to obtain the photon wavelength by remembering that $\nu = \frac{c}{\lambda}$, where c is the speed of light and λ is the wavelength.

Some examples of **hydrogenic** atoms and ions include: H, He^{+}, Li^{2+}, Be^{3+}, etc.

An **emission** spectrum is produced by electrons transitioning from excited (high energy) states to lower energy states. The energy lost by the electrons during such transitions is given out in the form of photons, which have an energy that corresponds to the difference in energy between the two states.

Hydrogenic emission spectra are produced by **hydrogenic** atoms or ions—species that have only a single electron. The presence of only one electron results in a greatly simplified spectrum when compared to the spectra of multi-electron atoms and ions.

In contrast to expectation from classical physics, analysis of atomic emission spectra shows that the emitted photons are only observed at discrete frequencies, rather than in a continuous spectrum. This provides direct experimental evidence for the fact that the electronic energy levels in atoms are **quantized** (i.e. they are fixed at discrete values). By convention, the energy levels are labelled in ascending order of energy by a **quantum number**, denoted by the letter n, which may take positive integer values ≥ 1. The energies of these levels are **negative**, because energy must be given to an electron to promote it to a higher energy level. This means that the smaller the value of n, the more tightly bound the electron. To a very good first approximation, the electronic energy levels and resultant emission spectra of **hydrogenic** atoms or ions can be described by this single quantum number. The requirement for and use of additional quantum numbers is covered in section 1.4.

The Rydberg equation

For hydrogenic atoms or ions, the empirically derived Rydberg equation can be used to calculate various properties of a photon emitted by the transition of an electron between an excited state to a lower-lying state.

One form of the Rydberg equation is:

$$\bar{\nu} = R_H Z^2 \left(\frac{1}{n_f^2} - \frac{1}{n_i^2} \right)$$

where $\bar{\nu}$ is the transition in **wavenumbers** ($\bar{\nu} = \frac{1}{\lambda}$), R_H is the Rydberg constant (units = wavenumbers), Z is the nuclear charge, and n_i and n_f are the principal quantum numbers for the initial (excited) and final (lower) energy levels respectively. For hydrogen, $Z^2 = 1$ so the formula is also commonly given as:

Keep a close eye on the units given for the Rydberg constant. The quantity calculated has the same units as the Rydberg constant used and as a result you may need to perform a unit conversion when giving your answer if a specific unit is requested. The numerical value of the constant will change depending on which units are used.

$$\bar{\nu} = R_H \left(\frac{1}{n_f^2} - \frac{1}{n_i^2} \right)$$

The form of the Rydberg equation given above calculates the wavenumber of the photons. As such, in order to determine other properties, we must be able to convert between these units. See section 1.1 for examples.

Worked example 1.3A

An important line in the emission spectrum of hydrogen is the so-called **H-alpha** (Hα) line, which is used to study features in the atmosphere of the sun. The Hα line is the first line in the **Balmer series** of hydrogen spectral lines. The lines in the Balmer series all result from transitions that end with $n_f = 2$.

(a) Calculate the wavelength of this line (give your answer in nm) and state where in the electromagnetic spectrum it lies.

(b) Though mainly composed of hydrogen, the sun also contains a large quantity of helium. Calculate the wavelength of the same transition for the hydrogenic ion He^+, and compare the value obtained to that for hydrogen. Is this a higher or lower energy transition than for hydrogen? Why?

The Rydberg constant, $R_H = 109737\ cm^{-1}$.

Solution

(a) This part is concerned with the hydrogen emission spectrum, specifically the Hα line, which we are told is the **first** line in the Balmer series of spectral lines, which features transitions that **end with $n = 2$.**

This may be decoded as follows: $n_f = 2$ and because this is the **first** line in the series and n_i must be an integer greater than n_f, then $n_i = (2+1) = 3$. We may now use these numbers in the Rydberg equation. The value for the Rydberg constant we are given ($R_H = 109737\ cm^{-1}$) means that our answer will also be in cm^{-1}. We have been asked to calculate the wavelength, and give our answer in nanometres, and as a result, we will need to perform unit conversions.

$$\bar{\nu} = R_H\left(\frac{1}{n_f^2} - \frac{1}{n_i^2}\right)$$

$$= 109737\ cm^{-1} \times \left(\frac{1}{2^2} - \frac{1}{3^2}\right)$$

$$= 109737\ cm^{-1} \times \left(\frac{1}{4} - \frac{1}{9}\right)$$

$$= 109737\ cm^{-1} \times \left(\frac{5}{36}\right)$$

$$= 15241\ cm^{-1}$$

$$\bar{\nu} = \frac{1}{\lambda}$$

$$\lambda = \frac{1}{\bar{\nu}} = \frac{1}{15241\ cm^{-1}} = 6.561\times10^{-5}\ cm = 6.561\times10^{-7}\ m = 656.1\ nm$$

Remember that $1\ cm = 10^{-2}\ m$, and $1\ nm = 10^{-9}\ m$.

A wavelength of 656.1 nanometres means that this line falls in the visible part of the electromagnetic spectrum, which ranges from approximately 390 to 700 nanometres. The Hα line is at the longer-wavelength end of the visible spectrum and as such is a deep red colour.

(b) This part asks us to calculate the wavelength for the same transition (i.e. $n = 3 \rightarrow n = 2$) for the hydrogenic ion, He^+. Since we are no longer dealing with hydrogen, we must extend the Rydberg equation to account for the additional nuclear charge of the He^+ ion. Helium has an atomic number of two, and therefore has a +2 nuclear charge.

$$\bar{\nu} = R_H Z^2\left(\frac{1}{n_f^2} - \frac{1}{n_i^2}\right) = 109737\ cm^{-1} \times 2^2 \times \left(\frac{1}{2^2} - \frac{1}{3^2}\right) = 109737\ cm^{-1} \times 4 \times \left(\frac{1}{4} - \frac{1}{9}\right)$$

$$= 60965\ cm^{-1}$$

$$\lambda = \frac{1}{\bar{\nu}} = \frac{1}{60965\ cm^{-1}} = 1.640\times10^{-5}\ cm = 1.640\times10^{-7}\ m = 164\ nm$$

A wavelength of 164 nm falls into the far ultra-violet part of the electromagnetic spectrum, and cannot be detected by the human eye. We can calculate the energy of a photon given its wavelength, by using the Planck relation, $E = h\nu = \frac{hc}{\lambda}$. It can be seen that the energy is inversely proportional to the wavelength. Therefore, a shorter wavelength has a higher energy and therefore the helium transition is higher in energy than the hydrogen transition.

This can be explained by the higher nuclear charge experienced by the electron in the He^+ ion. The higher nuclear charge (+2 vs +1 for hydrogen) means that the electron is more tightly bound to the nucleus, therefore requiring more energy to be promoted into a higher energy state.

Worked example 1.3B

(a) Calculate the frequencies of the first five lines in the **Lyman series** of lines in the hydrogen emission spectrum, which is comprised of transitions that end with $n_f = 1$. Give your answer in THz, to the nearest whole number. Plot these lines on an axis, and comment on any trends you see.

(b) Using the Rydberg equation, calculate the ionization energy of hydrogen giving your answer in kJ mol^{-1}.

The Rydberg constant, $R_H = 109737\ cm^{-1}$.

Solution

(a) Here, we are asked to calculate frequencies. However, the Rydberg constant is given to us in units of cm^{-1}. As such, we will need to perform a unit conversion.

The lines in the Lyman series all result from transitions that end with $n_f = 1$. Therefore, the Rydberg equation can be simplified as follows:

$$\bar{\nu} = R_H\left(\frac{1}{n_f^2} - \frac{1}{n_i^2}\right) = R_H\left(\frac{1}{1^2} - \frac{1}{n_i^2}\right) = R_H\left(1 - \frac{1}{n_i^2}\right)$$

The resultant wavenumbers may then be converted into wavelengths by noting that $\lambda = \frac{1}{\bar{\nu}}$. These wavelengths will be in units of centimetres, and so must be converted to metres. Once these values have been obtained, we can use $\nu = \frac{c}{\lambda}$ to obtain our frequencies, then convert to THz by noting that $1\ THz = 10^{12}\ Hz$.

This type of question is best answered using a table such as the example given in Table 1.1 to keep track of all steps used to obtain an answer. Spreadsheet software can be used to aid calculations and plotting.

Figure 1.1 shows the result of plotting these lines on an axis.

As can be seen, as the value of n_i (the initial excited state) increases, the difference in frequency between neighbouring spectral lines decreases, resulting in a convergent series of lines.

(b) The most difficult aspect of this part of the question is understanding how to use the Rydberg equation to account for ionization. Once this has been achieved, answering the rest of the question will only require simple unit conversions.

The most tightly bound state an electron in a hydrogen atom can experience is when $n = 1$. As the electron enters increasingly excited states, and n increases, the electron gets further from the nucleus and is thus less tightly bound.

Table 1.1 Steps in the solution to Worked example 1.3B part (a)

n_i	$1-\frac{1}{n_i^2}$	$\bar{\nu}$/cm^{-1}	λ/cm	λ/m	ν/Hz	ν/THz
2	$\frac{3}{4}$	82301	1.215×10^{-5}	1.215×10^{-7}	2.467×10^{15}	2467
3	$\frac{8}{9}$	97544	1.025×10^{-5}	1.025×10^{-7}	2.924×10^{15}	2924
4	$\frac{15}{16}$	102878	9.720×10^{-6}	9.720×10^{-8}	3.084×10^{15}	3084
5	$\frac{24}{25}$	105348	9.492×10^{-6}	9.492×10^{-8}	3.158×10^{15}	3158
6	$\frac{35}{36}$	106689	9.373×10^{-6}	9.373×10^{-8}	3.198×10^{15}	3198

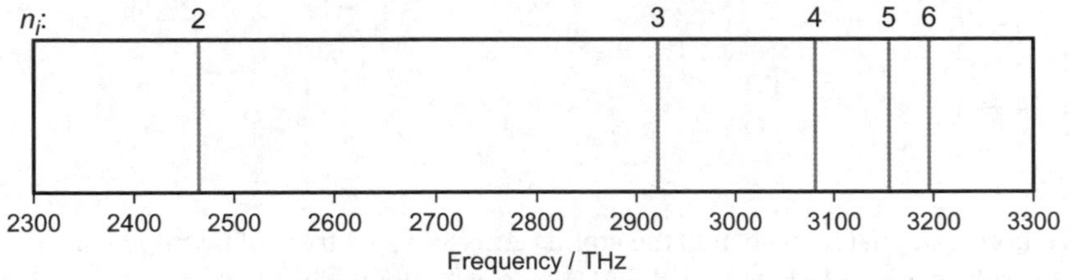

Figure 1.1 The frequencies of the first five lines seen in the **Lyman series** of hydrogen spectral lines.

If we define an electron as being ionized when $n_i = \infty$, then we can see how the Rydberg equation can be used to obtain the ionization energy for a single hydrogen atom.

$$\bar{\nu} = R_{\mathrm{H}}\left(\frac{1}{1^2}-\frac{1}{\infty^2}\right) = R_{\mathrm{H}}\left(1-\frac{1}{\infty}\right) = R_{\mathrm{H}}(1-0) = R_{\mathrm{H}} = 109737\ \mathrm{cm^{-1}}$$

Now we convert from cm^{-1} to m^{-1} in order to have all units in their correct SI forms.

$$1\ \mathrm{cm} = 0.01\ \mathrm{m}$$

$$(1\ \mathrm{cm})^{-1} = (0.01\ \mathrm{m})^{-1}$$

$$1\ \mathrm{cm^{-1}} = 100\ \mathrm{m^{-1}}$$

$$\bar{\nu} = R_{\mathrm{H}} = 10973700\ \mathrm{m^{-1}}$$

$$E_{\mathrm{photon}} = \frac{hc}{\lambda} = hc\bar{\nu} = 6.626\times10^{-34}\ \mathrm{J\,s}\times2.998\times10^{8}\ \mathrm{m\,s^{-1}}\times10973700\ \mathrm{m^{-1}}$$

$$= 2.180\times10^{-18}\ \mathrm{J}$$

Now we have the energy per photon (equivalent to the ionization energy), all that is left is to calculate the energy of a mole of photons then using the appropriate conversion factor (1 kJ = 1000 J) to obtain an answer in kJ mol^{-1}.

$$E_{\mathrm{mole}} = E_{\mathrm{photon}} \times N_{\mathrm{A}}$$

$$= 2.180\times10^{-18}\ \mathrm{J}\times6.022\times10^{23}\ \mathrm{mol^{-1}}$$

$$= 1.313\times10^{6}\ \mathrm{J\,mol^{-1}}$$

$$= 1313\ \mathrm{kJ\,mol^{-1}}$$

➔ That the series of lines converge on a single value should make sense when looking at the plot made in part (a). It can be seen that as n_i increases, the difference in frequency between neighbouring lines decreases, eventually converging on a single value—the ionization energy.

Question 1.5

(a) For the Paschen series of lines in the atomic spectrum of hydrogen, the final quantum number, $n_f = 3$. What are the frequencies of the first three lines in this series? (Rydberg constant, $R_H = 3.29 \times 10^{15}$ Hz)

(b) Would you expect the first three lines of the Balmer series to have higher or lower frequencies than the first three lines of the Paschen series? Explain your answer.

Question 1.6

(a) The highest energy series of lines in the atomic emission spectrum of hydrogen is known as the **Lyman** series; the next highest energy series is known as the **Balmer** series. Identify which of the following transitions belongs to the Lyman series and which to the Balmer series:

i. $3 \rightarrow 1$.
ii. $3 \rightarrow 2$.
iii. $4 \rightarrow 3$.
iv. $4 \rightarrow 2$.
v. $5 \rightarrow 1$.

(b) Demonstrate that lines seen in the atomic emission spectrum of hydrogen at 656 and 486 nm both belong to the Balmer series. The Rydberg constant, $R_H = 1.097 \times 10^7\ \text{m}^{-1}$.

1.4 Quantum numbers and atomic orbitals

Quantum numbers

The Schrödinger equation, published in 1926, provides a description of how a quantum mechanical system behaves and changes over time. It makes use of a mathematical description of the de Broglie waves (see section 1.1) associated with the system, which is known as the **wavefunction** for the system. The Schrödinger equation for the hydrogen atom results in the introduction of two new quantum numbers, which are denoted by the letters l and m_l, in addition to the quantum number n introduced previously.

The quantum number, n, is known as the principal quantum number. It may take integer values from 1 to ∞. The second quantum number, l, is known as the secondary quantum number (though is also commonly referred to as the angular, azimuthal, or orbital quantum number). It may take values from 0 to $(n-1)$. The third quantum number, m_l is known as the magnetic quantum number, and can take values from $-l$ to $+l$.

Due to the finite number of elements on the periodic table, in chemistry, generally we only encounter values of n ranging from $n = 1$ to $n = 7$ and values of l ranging from $l = 0$ to $l = 3$.

An **atomic orbital** is a single-electron wavefunction that describes a region of space in which an electron can be calculated to exist, and is described by a combination of these three numbers. Each quantum number provides useful information: n describes the energy of the orbital, l describes the shape of the orbital, and m_l describes the orientation of the orbital. It is clear from the definitions of allowed values above that different orbitals may share common values of n, or n and l. Orbitals with the same value of n are said to be in the same **shell**, while those with the same value of n and l are said to be in the same **subshell**.

Atomic orbitals in the same subshell have the same energy (i.e. are degenerate), and as such, we usually refer to atomic orbitals by their principal and secondary quantum numbers only. The principal quantum number is listed first as an integer, while the secondary quantum number is listed second and represented by a letter. For historical reasons, the letters are assigned as follows:

Value of *l*	**0**	**1**	**2**	**3**
Letter	*s*	*p*	*d*	*f*

Letters for values of l greater than 3 proceed in alphabetical order, starting with g for $l = 4$.

By way of example, orbitals with $n = 3$ and $l = 2$ are referred to as the 3*d* orbitals, while orbitals with $n = 2$ and $l = 1$ are referred to as the 2*p* orbitals.

Worked example 1.4A

(a) What are the possible values of m_l for the 2*p* orbitals? How many orbitals are in this subshell?

(b) How many subshells are possible when $n = 4$?

(c) How many atomic orbitals exist up to and including the shell where $n = 3$? Give the names of the subshells in your answer.

Solution

(a) The values allowed for m_l are integers that run from $-l$ to $+l$. A 2*p* orbital has $l = 1$ and therefore, the allowed values for m_l are:

$m_l = -1, 0, 1$

This implies that there are three **degenerate** orbitals in the 2*p* subshell.

Orbitals (or quantum states of an atom or molecule) that have the same energy are said to be **degenerate**.

(b) Subshells have the same values of n and l. Given that we have been asked about a single value for n, this question is essentially asking us how many values of l are possible when $n = 4$. l may take integer values running from 0 to $(n - 1)$. In this case then, it implies that: $l = 0, 1, 2, 3$ and therefore, the shell with $n = 4$ has four subshells.

(c) This question is best answered using a table to keep track of the allowed combinations of n, l, and m_l; see Table 1.2. Remember the rules for allowed values of l and m_l given a value for n, then start from the lowest value of n and complete the table.

When $n = 1$, l can only take the value 0. This is the 1*s* subshell. As m_l can only take values from $-l$ to $+l$, then m_l can also only take the value 0. This means there is one orbital possible in the 1*s* subshell.

When $n = 2$, l can take the values 0 and 1. As before, if l is 0, then we have the 2*s* subshell which contains one orbital. If $l = 1$, we have the 2*p* subshell. In this case, m_l can take integer values from $-l$ to $+l$, and therefore can take values of -1, 0, and $+1$ meaning there are three orbitals possible.

Table 1.2 Atomic orbitals and subshells up to $n = 3$

n	*l*	m_l	**Subshell**	**Number of orbitals**
1	0	0	1*s*	1
2	0	0	2*s*	1
	1	−1, 0, +1	2*p*	3
3	0	0	3*s*	1
	1	−1, 0, +1	3*p*	3
	2	−2, −1, 0, +1, +2	3*d*	5
			Total:	14

Notice how subshells with the same value of l contain the same number of orbitals, regardless of the value of n.

When $n = 3$, l can take the values 0, 1, and 2. If l is 0 then we have the 3*s* subshell, containing one orbital. If l is 1, then we have the 3*p* subshell containing three orbitals. If l is 2, then we have the 3*d* subshell. The allowed values of m_l are −2, −1, 0, +1, +2 and so the 3*d* subshell contains five orbitals.

Radial and angular wavefunctions

The **wavefunction** of an electron in a hydrogen atom is a mathematical function that allows us to understand its behaviour. It is dependent on the position of the electron relative to the atomic nucleus, and as such, while it can be used in terms of the familiar Cartesian coordinates *x*, *y*, and *z*, because atoms are spherical, it is more convenient to formulate the wavefunction in terms of **spherical coordinates**, which depend on a radius r from the origin, and two angles, denoted by the Greek letters θ (theta) and ϕ (phi). The natures of Cartesian and spherical coordinates are compared in Figure 1.2.

The wavefunction is represented by the Greek letter ψ.

When using spherical coordinates, the wavefunction for an electron in any given orbital, $\psi_{n,l,m}(r, \theta, \phi)$, can be separated into two components: a radial part, $R(r)$ that depends only on r, and an angular part, $Y(\theta,\phi)$ that depends on θ and ϕ.

Note that while the wavefunction can be negative, the probability of finding an electron at a given location is proportional to the square of the wavefunction and so will always be positive.

$$\psi_{n,l,m}(r, \theta, \phi) = R_{n,l}(r) \times Y_{l,m}(\theta,\phi)$$

The radial wavefunction, $R_{n,l}(r)$, depends on the quantum numbers n and l, and contains information about what happens to the wavefunction with increasing distance from the nucleus. Plots of $R(r)$ for different orbitals can be seen in Figure 1.3.

At certain values of r for some orbitals, the value of $R(r)$ is zero. Regions of space where the radial wavefunction is zero are called **radial nodes**. The number of radial nodes for a given orbital can be calculated using:

Radial nodes are sometimes referred to as *surface nodes*.

$$\text{Radial nodes} = n - l - 1$$

According to the **Born interpretation** of quantum mechanics, the square of the wavefunction is proportional to the probability of finding the electron in a small volume of space, dV. Therefore, by extending this argument to the volume of a very thin spherical shell at some distance r from the nucleus, we obtain a **radial distribution function**, which allows us to plot the probability of finding an electron at given distance from the nucleus.

$$RDF = 4\pi r^2 R(r)^2$$

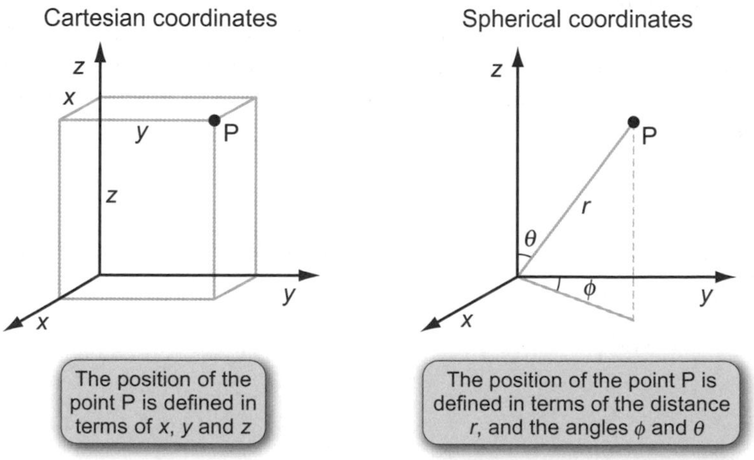

Figure 1.2 Cartesian and spherical coordinates both define the position of a point, P, with respect to an origin.

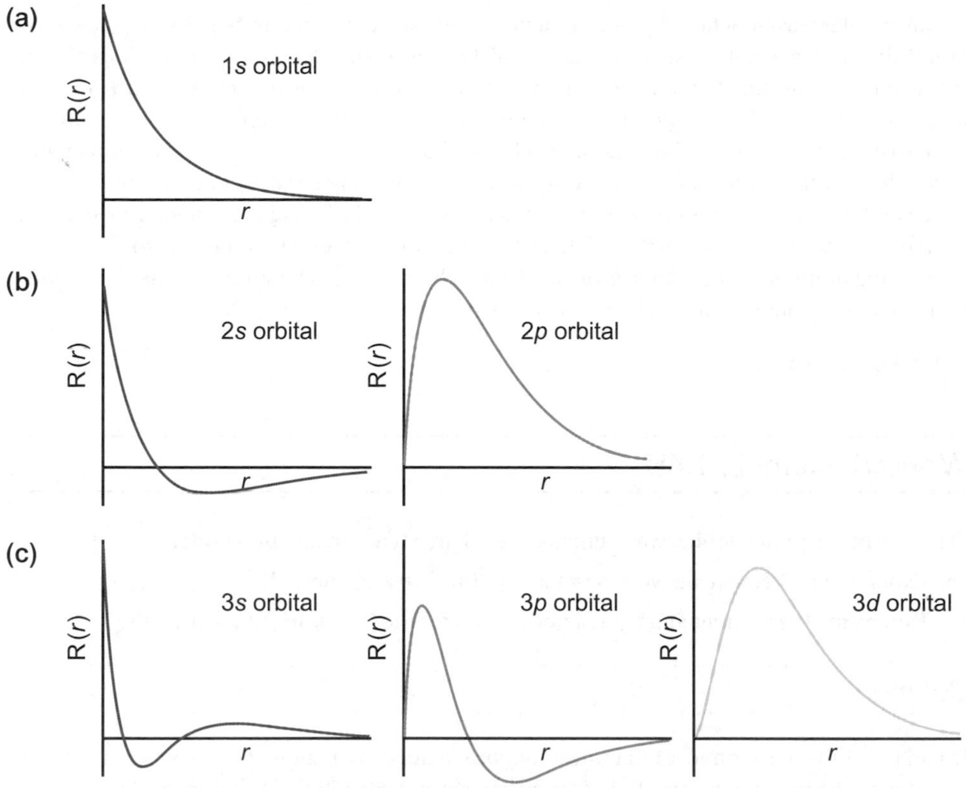

Figure 1.3 Variation of the radial wavefunction, *R*(*r*) with *r* for atomic orbitals in a hydrogen atom. (a) $n = 1$, (b) $n = 2$, (c) $n = 3$. Note: these plots are not all at the same scale.

These radial distribution functions also exhibit the radial nodes encountered earlier, which means that for certain orbitals, there are regions in space where an electron has zero probability of being found. It is also interesting to note that electrons in certain orbitals have a non-zero probability of being found inside the nucleus. The different behaviour of different types of orbital can help to explain a number of trends seen in the periodic table. Plots of radial distribution functions for various orbitals can be seen in Figure 1.4.

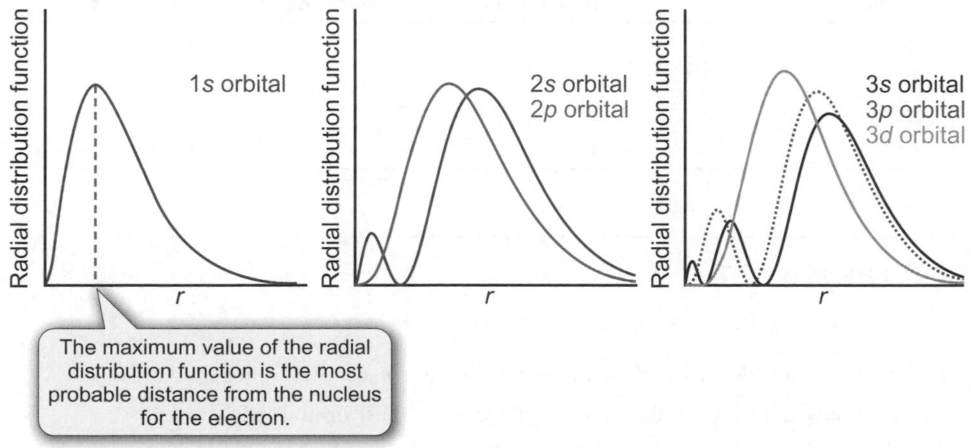

The maximum value of the radial distribution function is the most probable distance from the nucleus for the electron.

Figure 1.4 Variation of the radial distribution function with *r* for atomic orbitals with $n = 1$, 2, and 3 in a hydrogen atom.

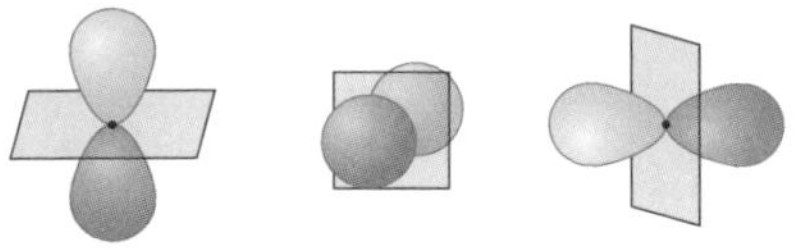

Figure 1.5 The three *p* orbitals each have one planar node.

Angular nodes are sometimes referred to as **planar nodes**.

Take care when using these relationships to determine the number of nodes, as it is easy to confuse the symbol 'l' for the number '1'. It may be simpler to think about the principal quantum number and the shapes of the orbitals when considering nodes.

The angular wavefunction, $Y_{l,m}(\theta, \phi)$, depends on the quantum numbers l and m_l, and contains information about the shape of the orbital. For certain orbitals and certain values of θ and ϕ, the angular wavefunction is zero. Regions where the angular wavefunction is zero are called **angular nodes**, which correspond to regions of space where there is zero probability of an electron existing. The number of angular nodes is equal to l. Therefore, an *s* orbital has no angular nodes, a *p* orbital has one angular node, a *d* orbital has two angular nodes and so on.

As can be seen in Figure 1.5, *p* orbitals which have $l = 1$, have angular nodes. This takes the form of a plane at the centre of the orbital, where the wavefunction is equal to zero.

The total number of nodes for a given orbital can be calculated by summing the radial nodes and the angular nodes, and can be calculated using:

Total nodes $= n - 1$

Worked example 1.4B

(a) In terms of atomic orbital wavefunctions, explain what is meant by a node.

(b) Explain the difference between a radial node and an angular node.

(c) Determine the number and type of nodes seen for the 2*s*, 3*p*, 4*d*, and 5*f* orbitals.

Solution

(a) In the context of atomic orbital wavefunctions, a node is a region of space where the value of the wavefunction is zero. Therefore, because the probability of locating an electron is proportional to the square of the wavefunction, at a node there is zero probability of locating an electron.

(b) A radial node results from the radial wavefunction being equal to zero. An angular node results from the angular wavefunction being equal to zero.

Plots of the radial distribution functions reveal that for a radial node, at a particular distance from the nucleus, regardless of the shape of the orbital, there is zero probability of locating an electron. In contrast to that, angular nodes form an integral part of the shapes of the orbitals, and manifest as **nodal planes** in the resultant orbital shape.

(c) The total number of nodes is given by $n-1$, the number of angular nodes is given by l, and the number of radial nodes is given by $n-l-1$.

Orbital	*n*	*l*	Total nodes	Angular nodes	Radial nodes
2*s*	2	0	1	0	1
3*p*	3	1	2	1	1
4*d*	4	2	3	2	1
5*f*	5	3	4	3	1

Question 1.7

Consider the orbitals 2*s*, 2*p*, 3*s*, 3*p*, and 3*d*.

(a) State the possible values of the quantum numbers *n*, *l*, and m_l for each of these.

(b) State how many nodes, and of what type, each of these orbitals has.

(c) Arrange these orbitals in order of ascending energy in a hydrogen atom.

Question 1.8

Identify the orbitals based on the number and type of nodes given in the table below.

	Radial nodes	Angular nodes
(a)	0	0
(b)	1	1
(c)	2	2
(d)	0	2
(e)	1	3

1.5 Multi-electron atoms and the periodic table

Orbital energies

The Schrödinger equation can only be solved **exactly** for atoms or ions with one electron. The so-called **orbital approximation** allows us to obtain very accurate information about the behaviour of multi-electron atoms and ions by assuming that the total wavefunction for an N-electron atom can be approximated by the product of N single-electron (i.e. hydrogenic) wavefunctions.

The energies of the orbitals in a hydrogenic atom or ion are dependent only on the principal quantum number, n (i.e. 2s and 2p orbitals have the same energy). However, the presence of multiple electrons removes this degeneracy: the energy of the orbitals now depends on the quantum numbers n and l, with higher values of both n and l corresponding to higher energies.

The orbital energies are as follows:

$1s < 2s < 2p < 3s < 3p < 4s < 3d < 4p < 5s < 4d < 5p < 6s < 4f \approx 5d < 6p < 7s < 5f \approx 6d$

The Aufbau principle, spin, and Hund's rule of maximum multiplicity

The **Aufbau principle** comes from the German word *aufbau*, which means 'building up'. It describes building up the electronic configuration of atoms by placing electrons into the lowest energy levels first. This naturally leads to the question of how many electrons can fit into an orbital.

Electrons have a quantum mechanical property called **spin**. Spin is an intrinsic property of subatomic particles, and can be thought of as the particle spinning around its axis, much like the earth rotates. Spin has an associated quantum number, m_s, which is known as the **spin magnetic quantum number**. For electrons, m_s can take two values: +½ or −½, which are often represented by ↑ and ↓ respectively. Therefore, to fully describe an electron in an atom, we must make use of four quantum numbers: the three quantum numbers encountered previously that describe the orbital, and the spin magnetic quantum number. The **Pauli exclusion principle** applied to electrons in atoms states that no two electrons can have the same four quantum numbers. This means that each orbital (where n, l, and m_l are fixed) can hold a maximum of two electrons.

To determine the **ground state** (i.e. lowest energy electronic configuration) of a particular atom, we also require information on how to place electrons into degenerate orbitals. **Hund's rule of maximum multiplicity** tells us that for degenerate orbitals, the lowest energy configuration has the maximum number of parallel electron spins. Therefore, according to this rule, of the three possible arrangements of two electrons filling a p subshell in Figure 1.6, option (a) is the most stable.

Note that these are generalized rules, and while they work for most of the elements on the periodic table, there are a number of important exceptions. Full explanations for those exceptions are outside the scope of this workbook. Section 3.6 of Burrows et al. (2013) has some examples, while others will be encountered in Chapter 5 of this workbook, which contains problems on the chemistry of the transition metals.

Electron configurations are written in the order in which the electrons are filled (i.e. lowest energy orbitals first), using superscript letters to state how many electrons are present in each orbital. To determine an electron configuration, we must know the number of electrons the atom or ion has.

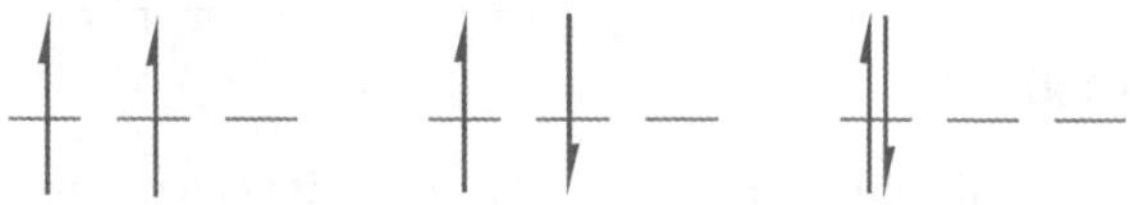

Figure 1.6 Three possible schemes for placing two electrons into a set of three degenerate orbitals. According to Hund's rule of maximum multiplicity, option (a) is the most stable.

The elements of the periodic table are arranged by their atomic number, which is the number of protons in the nucleus of each atom. Because a proton carries a +1 charge, for the atom to be neutral, there must be an equal number of negatively charged electrons surrounding it. We can then add or subtract electrons to account for a positively or negatively charged ion, and then use the Aufbau principle and Hund's rule of maximum multiplicity to fill the orbitals. If we know the electron configuration of the neutral atom, to form a cation we remove electrons from the highest energy orbitals first, while to form an anion we add additional electrons following the Aufbau principle.

A useful shorthand is to represent closed shells by the symbol of the noble gas that has that electronic configuration enclosed in square brackets. Some examples are given in Table 1.3.

Worked example 1.5A

Determine the ground state electronic configuration of the following atoms and ions and write it both in full and using the noble gas shorthand. State how many unpaired electrons each atom or ion has.

(a) Li

(b) B

(c) N

(d) O

(e) Si^+

(f) Cl^-

(g) Ca^{2+}

Solution

To answer this question, we must first determine how many electrons each atom or ion possesses. Once we have this number, we can use the Aufbau principle and Hund's rule of maximum multiplicity to fill the orbitals and determine the ground state electronic configuration of the elements.

(a) Lithium has an atomic number of three, and therefore the neutral atom must have three electrons. Using the Aufbau principle, we first fill the 1*s* orbital. The Pauli exclusion principle means that each orbital can contain only two electrons. The third

Table 1.3 Examples of the shorthand used to represent closed shells.

Element	Full electron configuration	Shorthand electron configuration
Li	$1s^2\ 2s^1$	[He] $2s^1$
F	$1s^2\ 2s^2\ 2p^5$	[He] $2s^2\ 2p^5$
K	$1s^2\ 2s^2\ 2p^6\ 3s^2\ 3p^6\ 4s^1$	[Ar] $4s^1$
O^{2-}	$1s^2\ 2s^2\ 2p^6$	[Ne]
Ca^{2+}	$1s^2\ 2s^2\ 2p^6\ 3s^2\ 3p^6$	[Ar]

electron must therefore occupy the next lowest energy orbital, which in this case is the 2*s* orbital:

i. Li: $1s^2\ 2s^1$.
ii. The shorthand [He] can be used to represent the $1s^2$ closed shell of electrons.
iii. Li: [He] $2s^1$.
iv. Neutral monatomic lithium therefore has one unpaired electron.

(b) Boron has an atomic number of five, and therefore we have five electrons in the neutral atom. Again, using the Aufbau principle and taking note of the Pauli exclusion principle, we fill the 1*s* and 2*s* orbitals encountered in part i. The next lowest energy orbital is the 2*p* orbital, and hence boron has the following electronic configuration:

i. B: $1s^2\ 2s^2\ 2p^1$.
ii. The shorthand [He] can be used to represent the $1s^2$ closed shell of electrons.
iii. B: [He] $2s^2\ 2p^1$.
iv. Neutral monatomic boron therefore has one unpaired electron.

(c) Nitrogen has an atomic number of seven, and hence has seven electrons to be accounted for in the neutral atom. As with boron, we fill the 1*s* and 2*s* orbitals leaving three electrons to distribute. As described in section 1.4, the 2*p* subshell is composed of three **degenerate** *p* orbitals, and hence the 2*p* orbitals can contain up to six electrons. Therefore, all of the remaining electrons can be contained in the 2*p* subshell:

i. N: $1s^2\ 2s^2\ 2p^3$.
ii. The shorthand [He] can be used to represent the $1s^2$ closed shell of electrons.
iii. N: [He] $2s^2\ 2p^3$.
iv. Hund's rule of maximum multiplicity tells us that the lowest energy configuration has the maximum number of parallel spins. Therefore, the ground state of a neutral nitrogen atom has three unpaired electrons, each occupying a 2*p* orbital.

(d) Oxygen has an atomic number of eight and as such the neutral atom has eight electrons. In light of our answer to part (c). and given that the 2*p* subshell can contain up to six electrons, we can determine that the ground state electronic configuration of oxygen is:

i. O: $1s^2\ 2s^2\ 2p^4$.
ii. The shorthand [He] can be used to represent the $1s^2$ closed shell of electrons.
iii. O: [He] $2s^2\ 2p^4$.
iv. Again, we use Hund's rule of maximum multiplicity to determine the number of unpaired electrons. Nitrogen features three unpaired electrons. Adding one electron to this means that we form one pair, and hence we see two unpaired electrons for neutral monatomic oxygen.

(e) Si has an atomic number of 14 and hence the neutral atom has 14 electrons to account for. We are looking at the ion Si^+ and hence we remove an electron to account for the single positive charge. The first ten electrons fully occupy the 1*s*, 2*s*, and 2*p* orbitals, and hence the remaining three electrons must occupy higher energy orbitals. The next lowest energy orbital available is the 3*s* orbital, which can hold up to two electrons, and hence this is fully filled. The final electron therefore occupies the next lowest energy level, which is the 3*p* subshell:

i. Si: $1s^2\ 2s^2\ 2p^6\ 3s^2\ 3p^1$.
ii. The shorthand [Ne] can be used to represent the $1s^2\ 2s^2\ 2p^6$ closed shell of electrons.
iii. Si: [Ne] $3s^2\ 3p^1$.
iv. This ion therefore has one unpaired electron.

(f) Cl has an atomic number of 17 and hence the neutral atom has 17 electrons to account for. In this ion, we have an additional electron to account for the single negative charge. Therefore, we must account for 18 electrons in total. As with silicon, the first ten electrons occupy the 1*s*, 2*s*, and 2*p* orbitals. Two more electrons occupy the 3*s* orbitals leaving six electrons to be accounted for. These can all be accommodated in the next lowest energy level, the 3*p* subshell:

i. Cl^-: $1s^2\ 2s^2\ 2p^6\ 3s^2\ 3p^6$.
ii. The shorthand [Ar] can be used to represent the $1s^2\ 2s^2\ 2p^6\ 3s^2\ 3p^6$ closed shell of electrons.
iii. Cl^-: [Ar].
iv. Cl^- has no unpaired electrons.

(g) Ca has an atomic number of 20 and hence the neutral atom has 20 electrons. We must therefore remove two electrons to account for the +2 charge on the ion, leaving 18 electrons to be accounted for. This is the same number seen for Cl^- and hence these two ions have the same electron configuration:

i. Ca^{2+}: $1s^2\,2s^2\,2p^6\,3s^2\,3p^6$.

ii. The shorthand [Ar] can be used to represent the $1s^2\,2s^2\,2p^6\,3s^2\,3p^6$ closed shell of electrons.

iii. Ca^{2+}: [Ar].

iv. Ca^{2+} does not have any unpaired electrons.

Turn to the Synoptic questions section on page 148 to attempt questions that encourage you to draw on concepts and problem-solving strategies from several topics within a given chapter to come to a final answer.

Final answers to numerical questions appear at the end of the book, and full worked solutions appear on the book's website. Go to http://www.oxfordtextbooks.co.uk/orc/chemworkbooks/.

Question 1.9

Write the full ground-state electron configurations of the following atoms and ions. Determine the number of unpaired electrons in each case.

(a) Na

(b) Na^+

(c) Cl

(d) Cl^-

(e) Al

(f) N^{3-}

(g) Mg

(h) Mg^{2+}

Question 1.10

Write the electron configurations of the following atoms and ions using the noble gas shorthand for a closed shell.

(a) Li

(b) S^{2-}

(c) Ca

(d) I^-

(e) Si

(f) K

(g) Ba^{2+}

(h) H^-

References

Burrows, A., Holman, J., Parsons, A., Pilling, G., and Price, G. (2013) *Chemistry*³, 2nd edn (Oxford University Press, Oxford).

2

Molecular orbitals and structure

2.1 Constructing molecular orbitals

Atomic orbitals (AOs, see Chapter 1) are single-electron wavefunctions that allow us to describe the behaviour of electrons in atoms. By extension, molecular orbitals are single-electron wavefunctions that describe the behaviour of electrons in molecules. Molecular orbital (MO) theory relies on two important approximations: the orbital approximation and the linear combination of atomic orbitals.

The **orbital approximation** is the assumption that the overall wavefunction for an atom containing *N* electrons can be approximated by the product of *N* single-electron atomic orbitals. Extending this idea to molecules leads us to the assumption that the overall wavefunction for an *N*-electron molecule can be approximated by the product of *N* single-electron molecular orbitals.

Wavefunctions have wave characteristics and therefore, atomic orbitals on one atom can interfere with those on another atom. The interference can be constructive (addition) or destructive (subtraction). We can therefore use an approach called the **linear combination of atomic orbitals** (LCAO) to build up molecular orbitals via addition and subtraction of atomic orbitals.

This can be expressed mathematically as follows:

$$\Psi = c_a\phi_a + c_b\phi_b$$

$$\Psi^* = c_a\phi_a - c_b\phi_b$$

Ψ and Ψ^* are the molecular bonding and antibonding orbitals respectively, ϕ_a and ϕ_b are the interacting atomic orbitals on atoms *a* and *b* respectively, and c_a and c_b are adjustable coefficients that can be positive or negative, that depend on the energy and symmetry of the interacting atomic orbitals on atoms *a* and *b* respectively.

In this chapter, we will use the Greek letter ϕ (phi) to represent atomic orbitals and the Greek letter Ψ (psi) to represent molecular orbitals. Some text books do not make this distinction, so pay close attention when studying from unfamiliar books.

Parity labels

As seen previously, atomic orbitals may have regions where the wavefunction is positive, and regions where it is negative. When we combine AOs to form MOs, attention should be paid to the symmetry of the resultant MO because there are important implications for spectroscopy.

To categorize MOs, we make use of the mathematical operation of **inversion**. This simply means taking a point at the coordinates (x, y, z) and mapping it onto a point at $(-x, -y, -z)$.

An orbital that appears identical after inversion is given the label *g*, derived from the German word *gerade*, which means even. An orbital that is not identical after inversion is given the letter *u*, from the German word *ungerade*, meaning odd.

Strength of the bonding

In complex molecules, we have multiple atomic orbitals overlapping to form multiple molecular orbitals. We can predict the strength of the bonding between the atoms on the basis of the number of electrons that occupy bonding orbitals versus the number of electrons that occupy antibonding orbitals.

To do this, we make use of the term **bond order**, which is defined as:

Bond order = (number of bonding pairs of electrons) – (number of antibonding pairs of electrons)

2.2 Molecular orbital energy level diagrams: homonuclear diatomics—overlap of *s* atomic orbitals to form molecular orbitals

As with atomic orbitals, the **Pauli exclusion principle** means that a molecular orbital can hold a maximum of two electrons. Therefore, using the Aufbau principle encountered in Chapter 1, we can construct an energy level diagram for molecules, which can provide useful insights into their properties and behaviour.

Figure 2.1 shows an example MO diagram for the simple homonuclear diatomic molecule H_2. As can be seen, the interaction of the two hydrogen 1*s* orbitals gives rise to two molecular orbitals: the $1\sigma_g$ bonding orbital that arises from constructive interference of the two atomic orbitals, and the $1\sigma_u{}^*$ antibonding orbital, which results from the atomic orbitals destructively interfering.

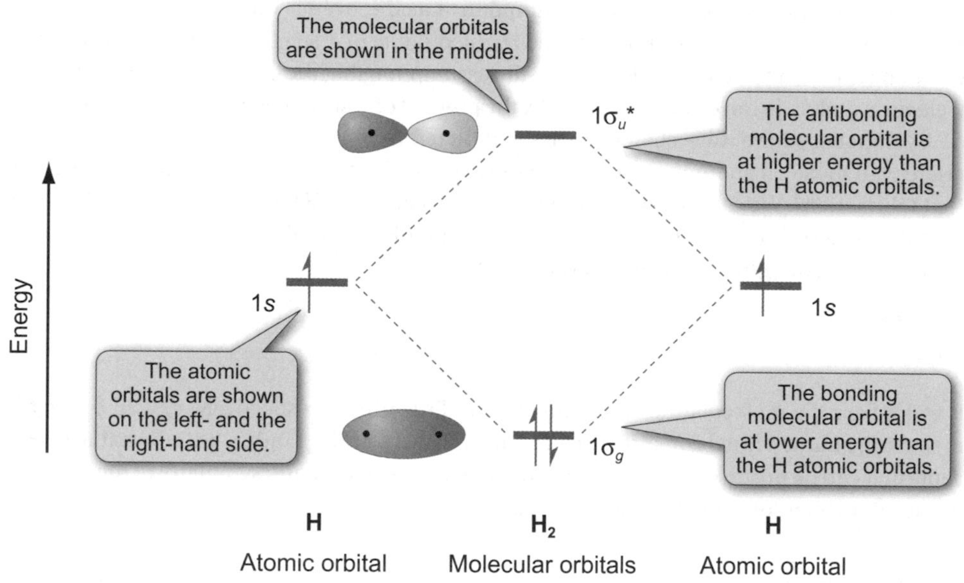

Figure 2.1 MO diagram for the simple homonuclear diatomic molecule, H_2.
Reproduced from Burrows et al., *Chemistry*[3] second edition (Oxford University Press, 2013). © Andrew Burrows, John Holman, Andrew Parsons, Gwen Pilling, and Gareth Price 2013.

Remember that a node is a region of space in which there is a zero probability of finding the electron. Nodes are discussed in more detail in Chapter 1.

The letter used to represent these orbitals, σ, is the Greek letter **sigma**.

Note that a given number of overlapping atomic orbitals will result in the same number of molecular orbitals. In Figure 2.1, two 1s AOs have combined to form two 1σ MOs.

As we shall see in later sections, a number of types of orbitals can interact to form a σ bond. The resultant σ bond in each case is cylindrically symmetrical (i.e. symmetrical with respect to rotation about the bond axis) and does not exhibit nodal planes that include the interatomic axis.

Figure 2.1 shows how the favourable constructive interference results in a decrease in energy of the bonding molecular orbital relative to its parent atomic orbitals, whilst the destructive interference results in an increase in energy for the antibonding orbital, relative to its parent atomic orbitals.

Worked example 2.2A

(a) According to the **linear combination of atomic orbitals** (LCAO) method for constructing molecular orbitals, which three criteria must be satisfied for a **bonding** molecular orbital to be formed?

(b) Sketch diagrams to show how *s* atomic orbitals may overlap to give bonding and antibonding σ molecular orbitals depending on the sign of the wavefunction in the overlapping lobes of the atomic orbitals.

(c) Explain how you would classify a molecular orbital as being:

i. bonding or antibonding.

ii. *gerade* (*g*) or *ungerade* (*u*).

(d) Sketch a molecular orbital energy level diagram to show the relative energies of the molecular orbitals in He_2. Use your diagram to predict whether or not you would expect this molecule to exist.

(e) Also use your diagram to predict whether or not the ion $[He_2]^+$ could exist.

Solution

(a) According to LCAO, three criteria must be satisfied for a **bonding** molecular orbital to be formed:

i. The orbitals must have the correct symmetry to overlap.

ii. The orbitals should be of similar energy so as to ensure efficient overlap.

iii. The overlap should occur in phase, i.e. the orbitals interact constructively.

(b)

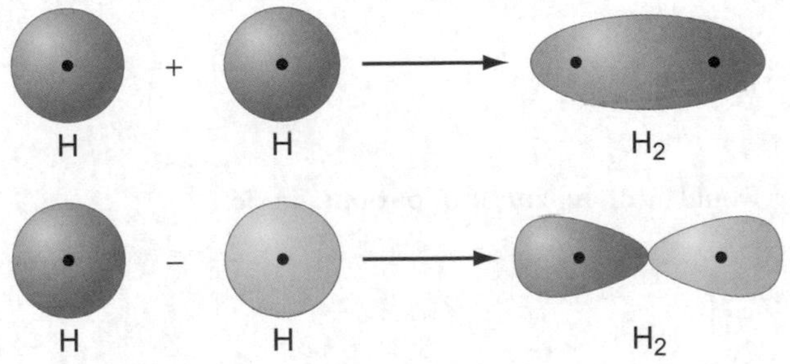

Reproduced from Burrows et al., *Chemistry*³ second edition (Oxford University Press, 2013). © Andrew Burrows, John Holman, Andrew Parsons, Gwen Pilling, and Gareth Price 2013.

(c) In order to classify orbitals correctly, we must look at their **symmetry**.

i. Bonding orbitals do not feature a nodal plane perpendicular to the internuclear axis. This essentially means that electron density is concentrated in between the atoms. Another way of looking at this is to see that bonding orbitals are symmetric with respect to rotation about an axis perpendicular to the internuclear axis. Antibonding orbitals do show a nodal plane, perpendicular to the internuclear axis, and are asymmetric with respect to rotation about an axis perpendicular to the internuclear axis.

ii. The *gerade* (*g*) and *ungerade* (*u*) parity labels refer to the symmetry of the orbital with respect to inversion, a mathematical operation that maps the point (x, y, z) onto the point $(-x, -y, -z)$. If the orbital is identical after inversion (i.e. keeps the same shape and phase), it is given the label *g*, which comes from the German word *gerade*, which

The orbitals of heteronuclear diatomic molecules are not given *g* and *u* labels. This is because the different sizes of the contributing atomic orbitals mean that inversion results in a changed shape and hence this cannot be described as *gerade* or *ungerade*.

means even. If the orbital keeps the same shape but changes its phase, then it is given the label *u*, which comes from the German word *ungerade*, which means odd.

(d)

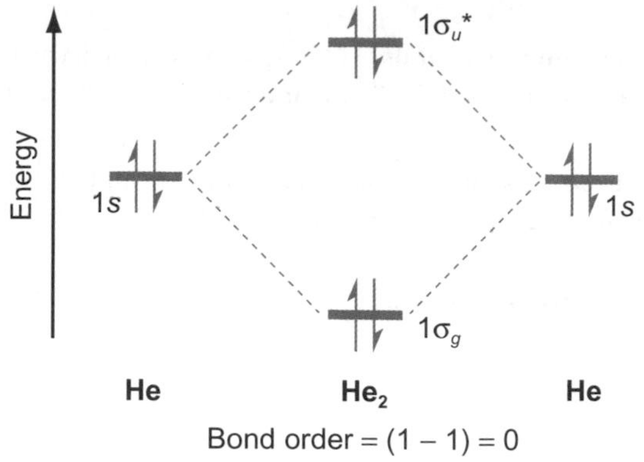

Reproduced from Burrows et al., *Chemistry*[3] second edition (Oxford University Press, 2013).

Because the number of bonding electrons is exactly matched by the number of antibonding electrons the molecule He_2 cannot exist as its bond order would be zero.

(e) In forming He_2^+ an electron is removed from the antibonding ($1\sigma_u{}^*$) orbital. Thus the overall bond order will be 0.5 which although indicating a weak bond also suggests that He_2^+ could, in principle, exist.

Question 2.1

Using the **Aufbau principle** and **Hund's rules of maximum multiplicity**, fill the following MO diagram with:

i. Four electrons.
ii. Six electrons.
iii. Eight electrons.

Decide if the resultant species would be **diamagnetic** or **paramagnetic**.

A species is said to be diamagnetic if all the electrons in the species are paired. A species is said to be paramagnetic if it contains unpaired electrons.

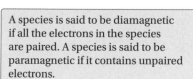

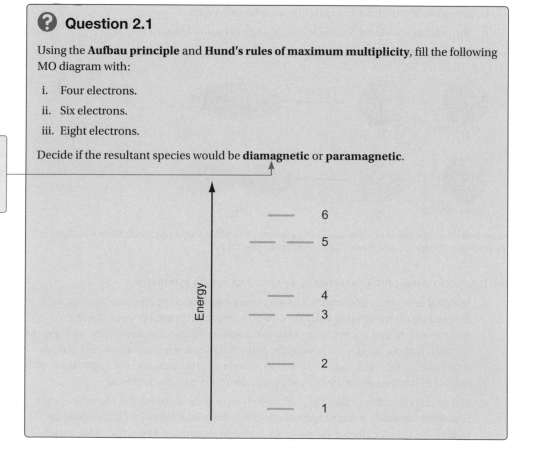

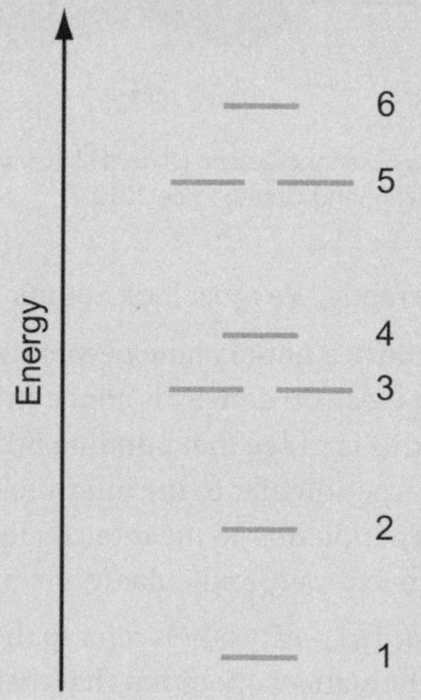

2.3 Molecular orbital energy level diagrams: homonuclear diatomics—overlap of *p* atomic orbitals to form molecular orbitals

When considering the overlap of p atomic orbitals it is important to consider the direction in which these orbitals are oriented. The standard convention is to label the interatomic axis of the molecule as z. Thus it may be seen that overlap of the p_z atomic orbitals will give rise to sigma bonding and antibonding molecular orbitals, as illustrated in Figure 2.2.

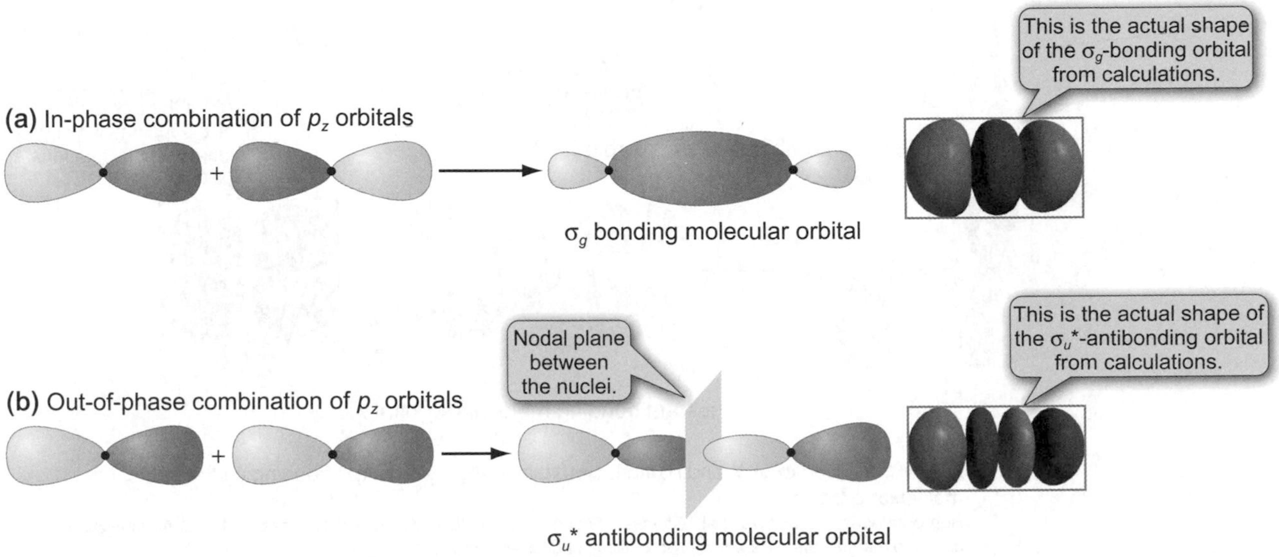

Figure 2.2 Overlap of p_z atomic orbitals giving rise to sigma bonding and antibonding molecular orbitals. Reproduced from Burrows et al., *Chemistry*[3] second edition (Oxford University Press, 2013). © Andrew Burrows, John Holman, Andrew Parsons, Gwen Pilling, and Gareth Price 2013.

By contrast the p_x and p_y atomic orbitals overlap to give π (pi) molecular orbitals. These orbitals are antisymmetric with respect to rotation about the interatomic axis, and feature a nodal plane which contains the interatomic axis.

We can understand that there are π orbitals if we consider the answer to Worked example 2.3A (part (a)).

Worked example 2.3A

(a) Sketch diagrams to show how atomic p orbitals may overlap to give either σ or π molecular orbitals depending on the orientation of the p orbitals. Show also how both bonding and antibonding orbitals may be produced.

(b) Draw a molecular orbital energy level diagram for the molecule O_2 showing overlap of the 2s and 2p orbitals using the linear combination of atomic orbitals (LCAO) method.

(c) Use your diagram to explain:

i. Whether O_2 is diamagnetic or paramagnetic.

ii. Whether the O—O bond would strengthen or weaken if O_2 lost an electron to form O_2^+.

(d) Use your diagram to decide whether O_2^- would have a stronger or weaker bond than O_2 and whether this ion would be diamagnetic or paramagnetic.

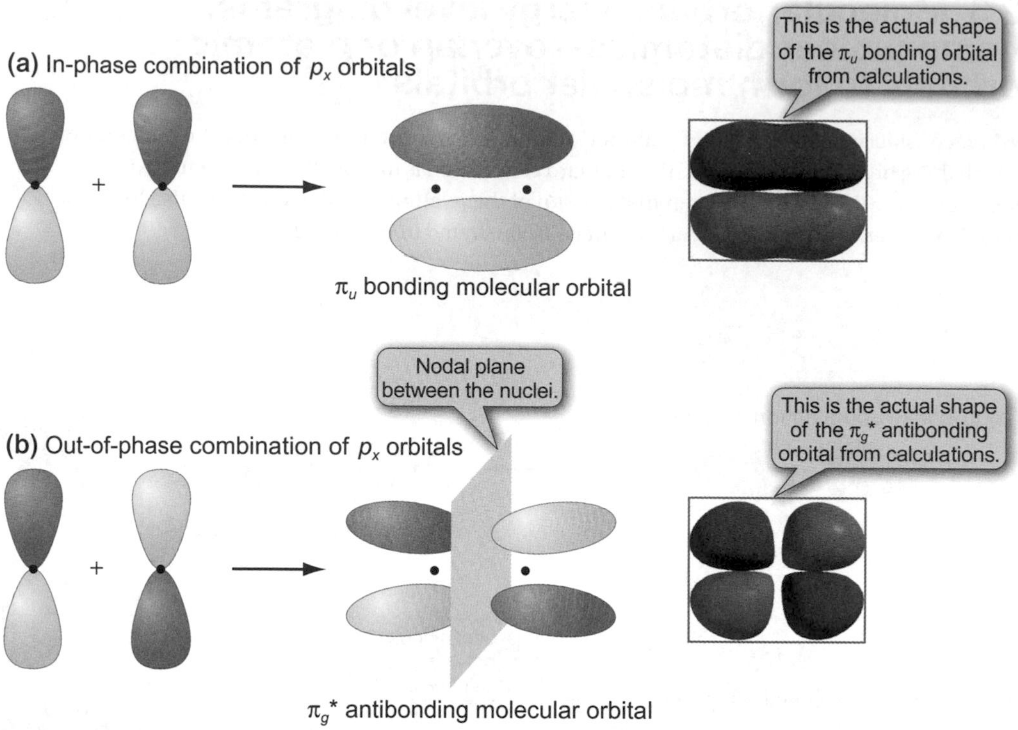

Figure 2.3 Overlap of p_x and p_y atomic orbitals giving rise to π (pi) bonding and antibonding molecular orbitals.

Solution

(a) Your diagrams should match Figures 2.2 and 2.3.

(b)

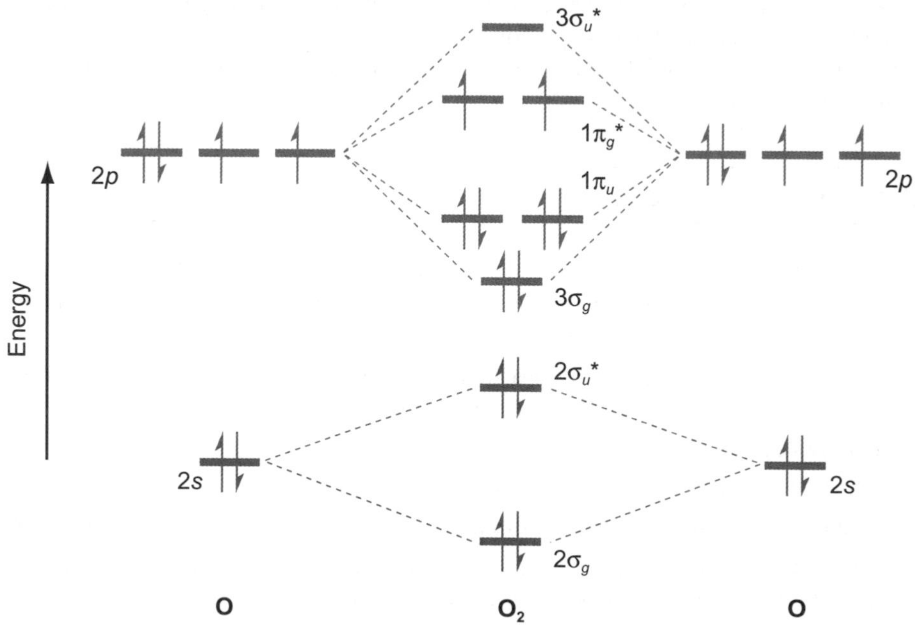

(c) If we consider the MO diagram in part (b) we can see that we have two unpaired electrons, occupying the $1\pi_g^*$ antibonding orbitals. (i) These unpaired electrons mean that oxygen is paramagnetic in the ground state. (ii) If an electron is lost from O_2 to form O_2^+ the electron will be lost from an antibonding orbital. As such the O—O bond will strengthen.

(d) O_2^- has one more electron than O_2. The $1\pi_g^*$ antibonding orbitals can accommodate a total of four electrons, and hence we expect the additional electron also to sit in the $1\pi_g^*$ antibonding orbitals; this would weaken the O—O bond. This would still leave one unpaired electron, and hence this ion is also paramagnetic.

Question 2.2

Draw a molecular orbital diagram for the heteronuclear diatomic molecule F_2 and use the diagram to answer the following questions:

(a) Is the molecule diamagnetic or paramagnetic?

(b) What is the bond order?

(c) Does the bonding get stronger, weaker, or remain unchanged when the molecule is ionized to form F_2^+?

(d) Does the bonding get stronger, weaker, or remain unchanged when the molecule is ionized to form F_2^-?

2.4 *s–p* mixing resulting in σ–π crossover

As we have seen in section 2.3, sigma bonding and antibonding molecular orbitals can be formed from the overlap of *s* and p_z atomic orbitals. If the *s* and *p* orbitals from which these sigma orbitals are formed are close together in energy, then **mixing** between the resulting sigma orbitals can occur. The result of mixing is that the higher energy orbital is raised in energy and the lower energy orbital is lowered in energy. If the mixing between the *s* and *p* orbitals is strong, then we may see the order of energies of the orbitals change. Taking the first row homonuclear diatomics as an example, if the 2*s* and 2*p* orbitals are relatively close in energy then mixing can occur and consequently $\sigma(2p_z)$ is raised in energy and $\sigma(2s)$ is lowered in energy. If the $\sigma(2p_z)$ orbital is raised in energy by a significant amount, it can move to a higher energy than the $\pi(2p_x)$ and $\pi(2p_y)$ orbitals. This is called **σ–π crossover**.

Worked example 2.4A

(a) Construct molecular orbital energy level diagrams for the homonuclear diatomic molecules B_2 and C_2.

(b) It is found experimentally that B_2 is paramagnetic and C_2 is diamagnetic. Explain why this information shows that σ–π crossover **must** have occurred in these molecules.

Solution

(a) In order to answer this question you need first to work out which molecular orbitals will be formed by overlap of 2*s* and 2*p* orbitals. The 2*s* orbitals will give σ (bonding) and σ* (antibonding) molecular orbitals. The $2p_z$ orbitals will likewise give σ(bonding) and σ* (antibonding) molecular orbitals. The $2p_x$ and $2p_y$ orbital will give π (bonding) and π* (antibonding) molecular orbitals. Now you need to decide on the sequence of these orbitals in terms of their energies. Because B and C lie to the left-hand side of the *p* block the *s*–*p* separation is small and so σ–π crossover occurs. Thus the $\pi(2p_{x,y})$ orbitals ($1\pi_g$) lie at **lower** energy than the $\sigma(2p_z)$ orbital ($3\sigma_g$).

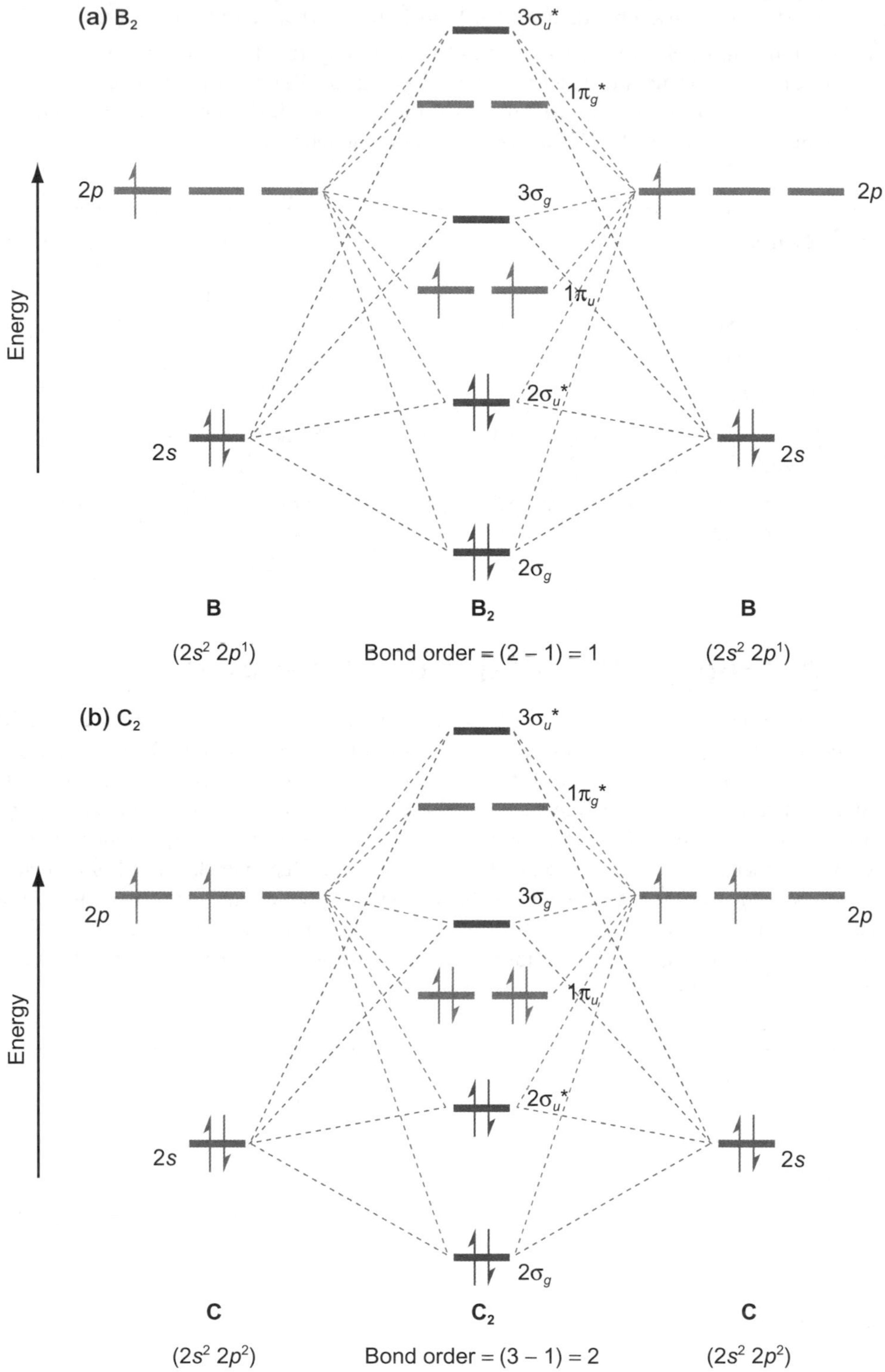

Reproduced from Burrows et al., *Chemistry*[3] second edition (Oxford University Press, 2013). © Andrew Burrows, John Holman, Andrew Parsons, Gwen Pilling, and Gareth Price 2013.

(b) If we consider the MO diagrams in (a) we can see that the HOMO (highest occupied molecular orbital) in these molecules are the $\pi(2p_{x,y})$ bonding molecular orbitals ($1\pi_g$), which because of σ–π crossover lie below the $\sigma(2p_z)$ orbitals ($3\sigma_g$) in energy. Four electrons may be accommodated in the $\pi(2p_{x,y})$ bonding molecular orbitals. B_2 has two electrons in these orbitals. From Hund's rule of maximum multiplicity (see Chapter 1) both orbitals will be half filled and the electrons will be unpaired. B_2 will therefore be **paramagnetic**. Two extra electrons are added in C_2. These will fill the $\pi(2p_{x,y})$ orbitals and so the molecule must be **diamagnetic**.

If we did not consider σ–π crossover when constructing our energy level diagram, and placed the $\sigma(2p_z)$ orbital lower in energy than the $\pi(2p_{x,y})$ orbitals, then we would expect B_2 to be diamagnetic and C_2 to be paramagnetic.

Worked example 2.4B

(a) Sketch a molecular orbital energy level diagram for the molecule Si_2.

(b) Experiments show that Si_2 is diamagnetic, i.e. has no unpaired electrons. How do you account for this observation using the diagram you have constructed in part (a)?

Solution

(a) This diagram will be almost identical to that for C_2 (see Worked example 2.4A) except that in Si_2 we must consider overlap of the 3*s* and 3*p* orbitals rather than the 2*s* and 2*p* orbitals which are considered in C_2.

(b) This argument is entirely analogous to the argument for C_2. The $\pi(3p_{x,y})$ bonding MO will be filled with four electrons arising from the 3*p* atomic orbitals of the two Si atoms.

Question 2.3

Would you expect to see more, or fewer, examples of σ–π crossover for the homonuclear diatomic molecules of the third row (Na_2–Cl_2) of the periodic table, than we see with the homonuclear diatomic molecules of the second row (Li_2–F_2)?

► **Hint** In your answer, consider how the difference in energy between adjacent orbitals changes as we go down the periodic table.

Question 2.4

Explain why σ–π crossover is observed for N_2 but not for F_2. Would you expect σ–π crossover to be more likely in Cl_2 than in F_2? Explain your answer.

2.5 Heteronuclear diatomics—σ bonding only

When considering **heteronuclear** diatomics we must return to the basic criteria of the **LCAO** theory for a bonding molecular orbital to be formed. In particular we must consider the relative energies of the orbitals on the two atoms and the area of overlap of the two orbitals. In terms of energies, the atomic orbital with the lower energy will contribute more to the bonding molecular orbital. This may be expressed mathematically by the following equations:

$$\Psi = c_a\phi_a + c_b\phi_b$$

$$\Psi^* = c_a\phi_a - c_b\phi_b$$

If the energy of orbital ϕ_a is lower than the energy of orbital ϕ_b, c_a will be greater than c_b. This implies that atom a is more **electronegative** than atom b. This should make sense from our understanding of atomic orbitals—the more electronegative the element, the higher the effective nuclear charge, and the more tightly the valence electrons are bound. (See Chapter 3).

This is expressed in the hypothetical molecular orbital diagram that describes the bonding between heteroatoms a and b, which is depicted in Figure 2.4.

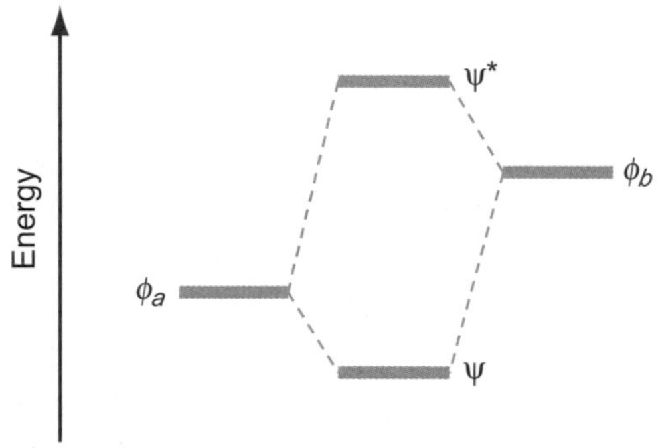

Figure 2.4 Hypothetical molecular orbital diagram that describes the bonding between heteroatoms *a* and *b*.

Reproduced from Burrows et al., *Chemistry*[3] second edition (Oxford University Press, 2013). © Andrew Burrows, John Holman, Andrew Parsons, Gwen Pilling, and Gareth Price 2013.

Atomic orbital ϕ_a is lower in energy than ϕ_b and hence contributes more to the resultant bonding molecular orbital, Ψ. The higher energy ϕ_b therefore contributes more to the resultant antibonding orbital Ψ^*.

The greater the difference in energy and size between the orbitals, the poorer the overlap and the weaker the net bonding interaction.

Worked example 2.5A

(a) Construct a molecular orbital energy level diagram for the molecule LiH.

(b) Explain, using your diagram, whether the bonding pair of electrons is more closely associated with the H or the Li atom.

(c) Explain whether the Li—H bond would become weaker or stronger if an electron were removed from the molecule to form $[LiH]^+$.

Solution

(a) In this case you should understand that the 1*s* orbital on H will overlap with the 2*s* orbital on Li to form σ (bonding) and σ* (antibonding) molecular orbitals and that there are two valence electrons that can fill the bonding orbital.

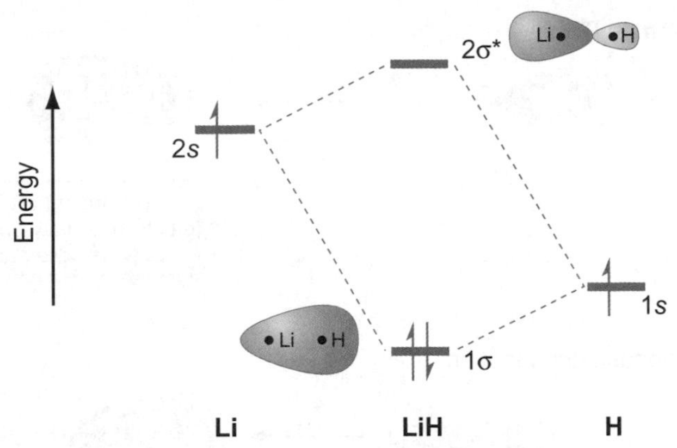

(b) We can see from this diagram that, as should be expected, the energy of the 1*s* orbital on the hydrogen atom is lower than that of the 2*s* orbital on the lithium atom. The hydrogen atomic orbital therefore contributes more to the resultant bonding molecular orbital, and as such, the bonding pair of electrons will be more closely associated with the hydrogen atom.

(c) Removing an electron from LiH, forming the ion $[LiH]^+$, would result in an electron being removed from a bonding orbital. This reduces the overall bond order from 1 to 0.5, and hence reduces the strength of the bond.

Overlap of *s* and *p* orbitals

The next point to consider is the area (and wavefunction sign) of the overlapping atomic orbitals. This may be considered in a situation where a sigma-bonded heteronuclear diatomic is formed by overlap of *s* and *p* orbitals. Figures 2.5 and 2.6 show that for the molecule HF a p_z orbital may overlap with an *s* orbital to form bonding and antibonding σ molecular orbitals; p_x and p_y cannot overlap in this way. In these cases the overlap is said to be **non-bonding** and does not contribute to the overall bonding of the molecule. Instead, these occupied orbitals form two of the three fluorine lone pairs of electrons. The hydrogen 1*s* and fluorine 2*s* orbitals are too far apart in energy to overlap effectively and so the fluorine 2*s* electrons form the remaining lone pair of electrons on the fluorine atom.

Worked example 2.5B

(a) Construct a molecular orbital energy level diagram for the molecule HF.

(b) Explain using your diagram whether the bonding pair of electrons is more closely associated with the H or F atom.

(c) Explain the observation that the bond strength of this molecule does not change significantly when the molecule is ionized to form $[HF]^+$.

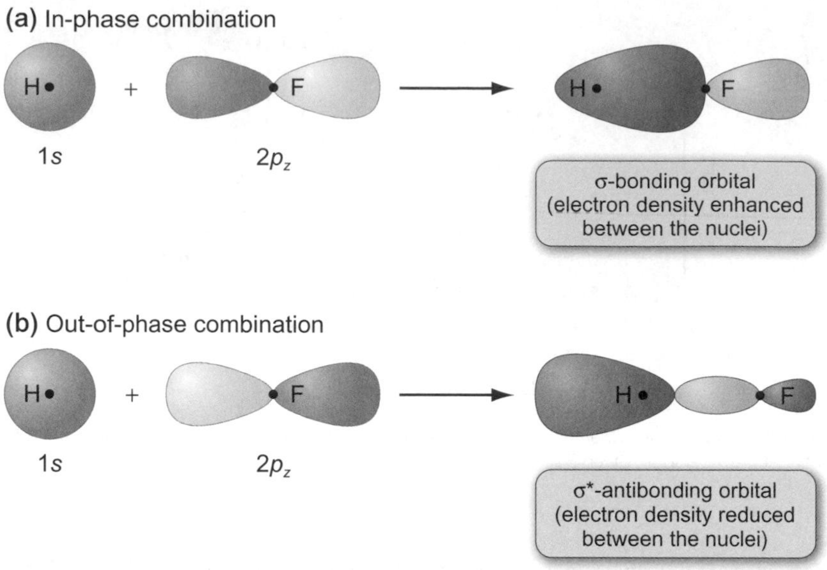

Figure 2.5 The interaction between a hydrogen 1s orbital and a fluorine $2p_z$ orbital leads to (a) a σ-bonding orbital and (b) a σ*-antibonding orbital.

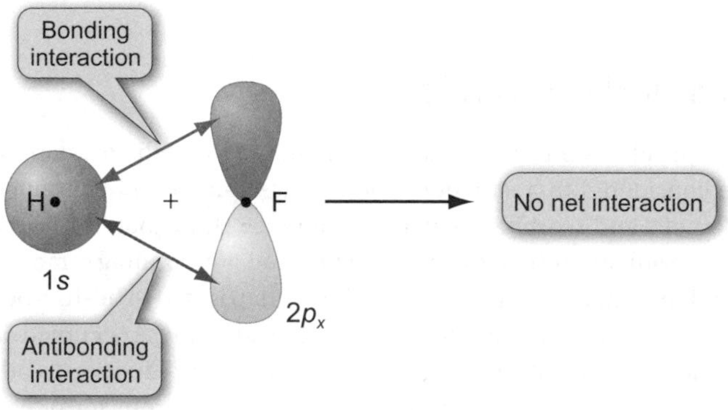

Figure 2.6 There is no net interaction between the H(1*s*) and F($2p_{x,y}$) orbitals.

Solution

(a) In this case the hydrogen 1*s* orbital will overlap with the $2p_z$ orbital on fluorine to form σ (bonding) and σ* (antibonding) molecular orbitals.

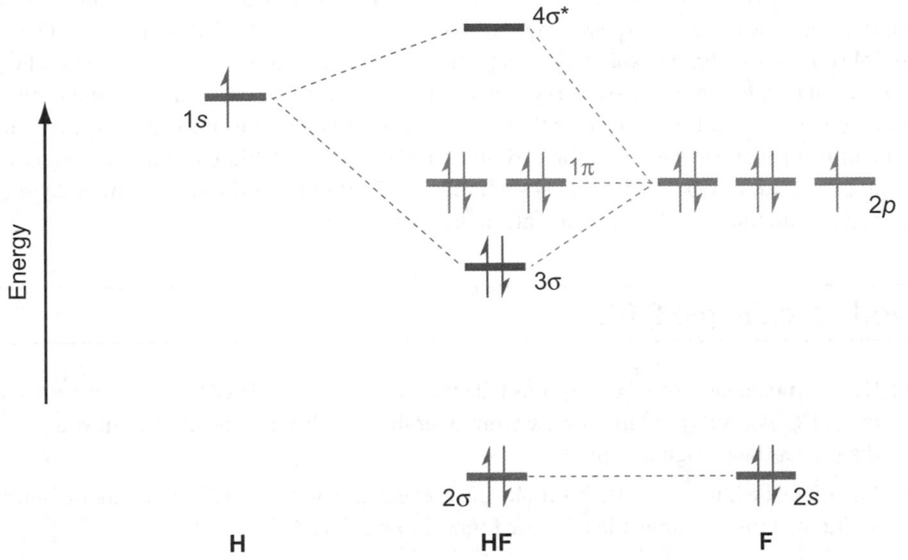

Reproduced from Burrows et al., *Chemistry*[3] second edition (Oxford University Press, 2013). © Andrew Burrows, John Holman, Andrew Parsons, Gwen Pilling, and Gareth Price 2013.

(b) Inspection of the MO diagram in the answer to part (a) shows that in HF the F $(2p_z)$ orbital lies at **lower** energy than the H(1*s*) orbital. Thus the bonding pair of electrons is more closely associated with the F atom. This is in line with electronegativity values where F is more electronegative than H (Pauling electronegativity: F = 3.98, H = 2.20).

(c) In part (b) we have discussed the bonding MO and bonding electrons. However, the HOMO in HF is **non-bonding**. These are the $2p_{x,y}$ orbitals on F which as seen previously, cannot overlap with H(1*s*) to form bonding or antibonding orbitals. Because the highest energy electrons are non-bonding then ionization to form the $[HF]^+$ cation will not lead to any significant change in bond strength.

Question 2.5

(a) Draw a molecular orbital energy level diagram for sodium hydride. Would you expect the bonding to be weaker or stronger than in lithium hydride?

(b) Decide if the following statements regarding heteronuclear diatomic molecules are true or false, and explain your answers:

i. All have bond orders greater than zero.

ii. The antibonding molecular orbitals reside more on the more electropositive element than on the more electronegative element.

iii. The greater the difference in energy between two orbitals, the better the overlap.

iv. The bonding molecular orbitals reside more on the more electropositive element than on the more electronegative element.

v. Their molecular orbitals are less symmetrical than those of homonuclear diatomic molecules.

2.6 Heteronuclear diatomics—σ and π bonding

The situation becomes more complex when we have heteronuclear diatomic molecules, in which both σ and π molecular orbitals are present. A good example is the molecule CO. In CO the O atom is more electronegative, so we may reasonably suppose that the electron pair in the CO σ bond (remembering that the C≡O triple bond is composed of one σ bond and two π bonds) would be more closely associated with the O atom. However, the bonding in CO is more complex than this simple explanation suggests. The interaction between the σ(2s) and $\sigma(2p_z)$ orbitals that leads to σ–π crossover is particularly pronounced for heteronuclear diatomic molecules (such as CO). This is because there tends to be a relative closeness in energy of some of the atomic orbitals (in this case, the carbon 2*s* and oxygen 2*p* orbitals), which allows efficient mixing to occur. In CO, this mixing is such that the 5σ bonding orbital has a high degree of *s* character, with the orbital located mainly on the carbon atom.

Worked example 2.6A

(a) Draw a molecular orbital energy level diagram for the molecule CO. You should consider orbital 'crossover' (σ–π) in drawing your diagram. Clearly label the orbitals in your diagram as being sigma or pi.

(b) Calculate the bond order in this molecule and explain what would happen to the bond strength of the CO molecule if it were ionized to produce CO^+.

Solution

(a)

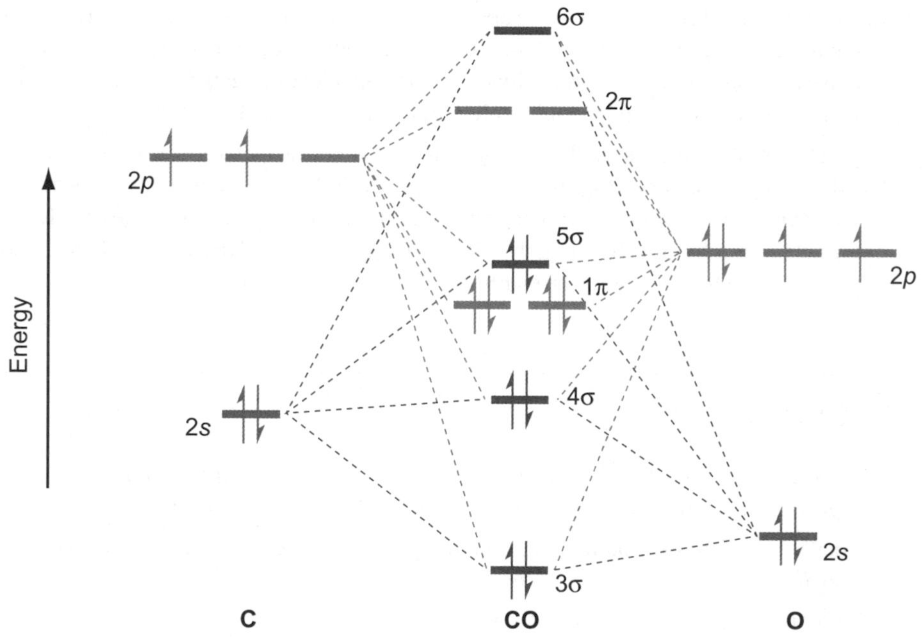

While O_2 does not show σ–π crossover, CO does. The relative energies of the 2*s* and 2*p* orbitals are suitably close for significant overlap to occur between the C(2*s*) and $O(2p_z)$ orbitals. As a result, we predict that σ–π crossover will be observed.

(b) There are four pairs of electrons occupying bonding orbitals (3σ, 1π, and 5σ) and one pair of electrons in an antibonding orbital (4σ), and hence carbon monoxide has a bond order of 3. As can be seen in the molecular orbital diagram in part (a), the HOMO is the 5σ bonding orbital, and hence removal of an electron results in a decrease in bond order (from 3 to 2.5) and hence the bond strength decreases.

Question 2.6

(a) Draw a molecular orbital energy level diagram for the molecule NO. You should consider orbital 'crossover' (σ–π) in drawing your diagram. Clearly label the orbitals in your diagram as being sigma or pi.

(b) Calculate the bond order in this molecule. Would you expect the bond to get longer, shorter, or remain the same if NO were ionized to produce (i) NO^+ (ii) NO^- ?

2.7 Triatomic molecules

When we come to consider triatomic molecules the situation becomes even more complicated: we now have to consider the overlap of three atomic orbitals. From the rule mentioned in section 2.2 that the number of **molecular** orbitals obtained is equal to the number of **atomic** orbitals that overlap, we can see that for every three overlapping atomic orbitals in a triatomic molecule, three molecular orbitals must be produced. These are characterized as bonding, antibonding, and **non-bonding** molecular orbitals. It is most straightforward to begin by considering a triatomic molecule (in this case a molecular ion) that has σ bonding only. This is considered in Worked example 2.7A.

Worked example 2.7A

(a) Draw diagrams to show how the H(1*s*) and F(2p_z) orbitals may overlap to produced molecular orbitals in the molecular ion $[HF_2]^-$.

(b) Sketch a molecular orbital energy level diagram for this ion showing the occupancy of the molecular orbitals.

(c) What is the bond order in this ion?

(d) State whether or not you would expect the bonds to get stronger, weaker, or remain at about the same strength if $[HF_2]^-$ were compared with the theoretical neutral molecule HF_2.

Solution

(a) The H(1*s*) orbital does not overlap effectively with the F(2*s*) orbitals due to the large difference in energy. Therefore, the molecular orbitals in this ion are formed from the overlap between the H(1*s*) orbital and the F(2p_z) orbitals. Figure 2.7 shows how the orbitals of these three atoms may interact to form bonding, non-bonding, and antibonding orbitals. Note that the non-bonding orbitals result from the 2p_z orbitals interacting out of phase. In this case, whatever the phase of the central 1*s* orbital, the result will be the same: one bonding interaction and one antibonding interaction and hence an overall non-bonding interaction.

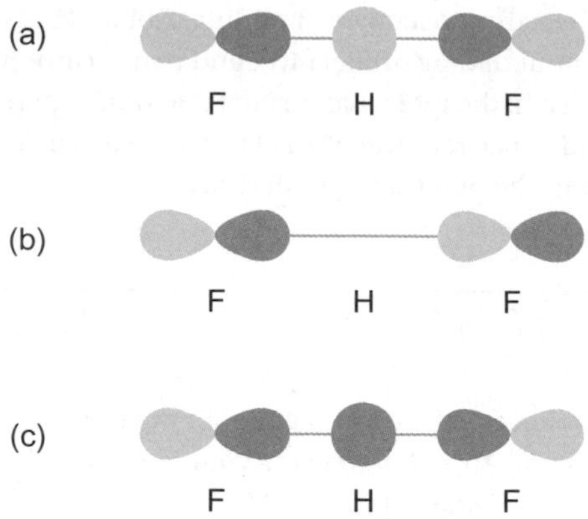

Figure 2.7 Schematic to show how overlap of the H(1*s*) and F(2p_z) orbitals can give rise to (i) anti bonding, (ii) non-bonding, and (iii) bonding molecular orbitals in the ion $[HF_2]^-$.

(b) Fluorine has seven valence electrons. If we assume that the extra electron associated with the negative charge on the ion resides on one of the fluorine atoms (a reasonable assumption given the high electronegativity of fluorine), then we must assume that one F(2p_z) orbital is full, and one is half filled. This gives a total of three electrons from the two fluorine atoms and one electron from the hydrogen atom, as shown in Figure 2.8.

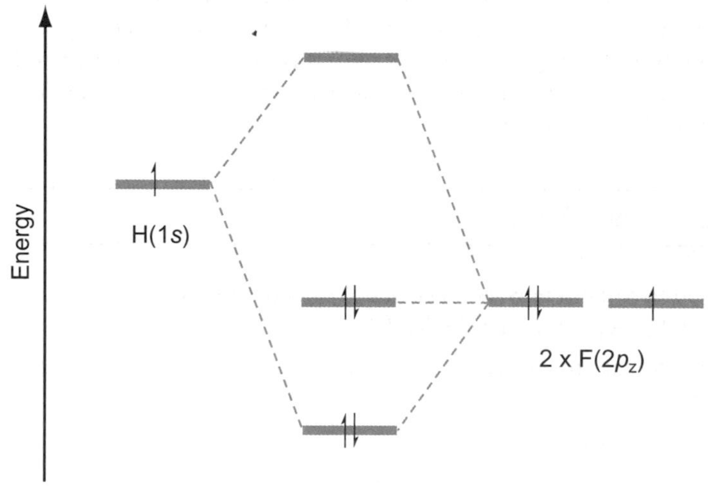

Figure 2.8 Molecular orbital energy level diagram showing the ground state electronic configuration in the ion $[HF_2]^-$.

(c) There are a total of four electrons—one from the H 1*s* orbital and three from the F 2p_z orbitals that must fill the molecular orbitals. Two of these occupy the bonding MO and two the non-bonding MO. The antibonding MO is vacant. Thus two bonding electrons hold together three atoms and this is known as a three-centre, two-electron bond. The bond order will be approximately 1/2, i.e. weaker than a single bond.

(d) If HF_2 could be made then the extra electron would be lost from a non-bonding molecular orbital and so one would expect the H–F bonds to remain at about the same strength.

Question 2.7

(a) Draw diagrams to show how the 2*s* and $2p_z$ atomic orbitals on beryllium may form bonding and antibonding interactions with the 1*s* orbitals on the H atoms in the molecule BeH_2.

(b) Construct a molecular orbital diagram for BeH_2 showing the occupancy of the molecular orbitals with electrons.

(c) Discuss whether or not you would expect ionization of BeH_2 to form $[BeH_2]^+$ to lead to any significant change in the bond lengths within the molecule.

BeH_2 exists as a molecular species in the gas phase. In the solid phase it forms a polymeric structure with Be—H—Be bridges. This structure is described in Chapter 3.

Question 2.8

(a) Sketch the shape of the octahedral molecule SF_6.

(b) SF_6 can be considered as containing linear SF_2 units. Sketch diagrams to show how the $2p_z$ orbitals (those lying along the SF_2 axis) on F and the $3p_z$ orbitals on S may overlap to give bonding, non-bonding, and antibonding interactions.

(c) Hence sketch a molecular orbital diagram for the SF_2 unit, showing occupancy of the molecular orbitals with electrons.

2.8 Electron-deficient molecules: hydrogen bridges

Three-centre, two-electron (3c2e) bonds are also used to describe the bonding in many **electron-deficient** molecules. In section 3.3, the structure of covalent beryllium hydride (BeH_2) is discussed. In the solid state, beryllium hydride exists as a giant covalent structure, with four-coordinate tetrahedral beryllium atoms joined together by hydrogen bridges, as illustrated in Figure 2.9. These bridges are good examples of 3c2e bonds.

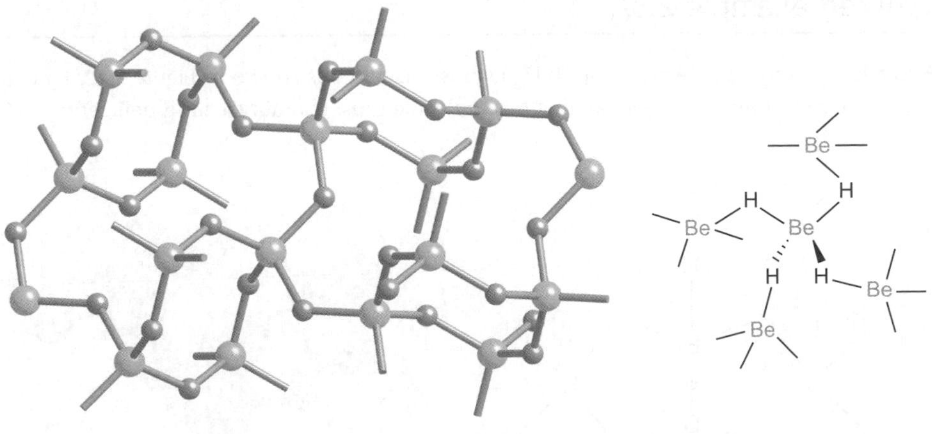

Figure 2.9 Solid BeH_2 forms a three-dimensional polymeric structure with bridging hydrogen atoms. Reproduced from Burrows et al., *Chemistry*[3] second edition (Oxford University Press, 2013). © Andrew Burrows, John Holman, Andrew Parsons, Gwen Pilling, and Gareth Price 2013.

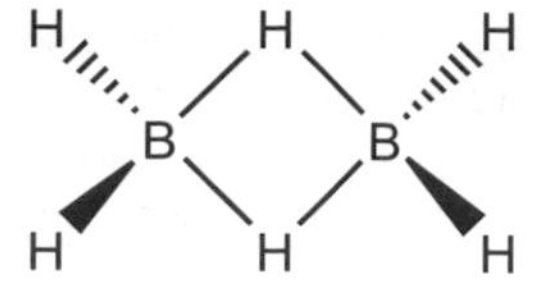

Figure 2.10 Diagram of the molecule diborane, which features hydride bridges.

Similar bonding is found in the molecule diborane (B_2H_6), as shown in Figures 2.10 and 2.11.

It is possible to show that this molecule is **electron deficient** by counting the number of valence electrons available, and comparing this number to the number of electrons that would be required for standard two-centre, two-electron (2c2e) bonds. In the B_2H_6 molecule there are a total of 12 valence electrons: three from each of the two boron atoms and one from each of the six hydrogen atoms. The structure shown in Figure 2.11 suggests that we have eight covalent bonds present. Eight standard 2c2e bonds would require 16 electrons, and hence this structure must be electron deficient. When the lengths of the terminal and bridging B—H bonds are examined, we observe that the bridging B—H bonds are longer than the terminal bonds, which provides physical evidence that the bonding is different for the two types of B—H bond.

Inspection of the structure illustrated in Figure 2.11 shows that eight of these electrons are used for the four terminal B—H bonds, each of which is a standard two-centre, two-electron (2c2e) covalent bond. This leaves just four electrons for the two B—H—B bridges and hence each of the B—H—B bridges must feature an electron deficient 3c2e bond.

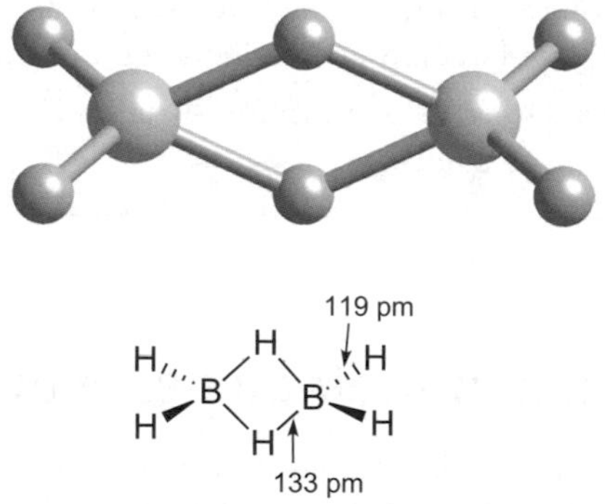

Figure 2.11 The structure of diborane (B_2H_6).
Reproduced from Burrows et al., *Chemistry*[3] second edition (Oxford University Press, 2013).

Worked example 2.8A

Consider the molecule tetraborane (B_4H_{10}) whose structure is shown in Figure 2.12. By counting valence electrons, demonstrate that this molecule must contain electron deficient B—H—B bridges.

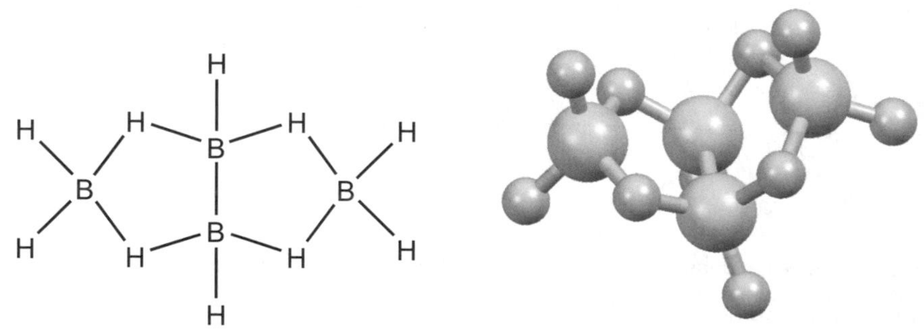

Figure 2.12 The structure of tetraborane, B_4H_{10}.

Solution

B_4H_{10} contains a total of 22 valence electrons (three from each of the four B atoms and one from each of the ten H atoms). The structure in Figure 2.12 shows 15 covalent bonds in total, which would require 30 electrons, and hence this structure is electron deficient. Of the 22 available valence electrons, 12 are used for the six B—H terminal bonds and two for the B—B bond. This leaves eight electrons for the four B—H—B bridges which must therefore be three-centre, two-electron bonds.

Valence electrons are the outermost electrons in an atom that are available for bonding. Hydrogen, of course, has just one valence electron ($1s^1$); boron has three valence electrons ($2s^22p^1$).

Question 2.9

State which of the following molecules or ions contain an electron-deficient bond.

(a) H_2

(b) $[H_2]^+$

(c) $[O_2]^{2+}$

(d) Li_2

(e) LiH

(f) BeH

(g) $[LiH]^+$

(h) $[F_2]^-$

Question 2.10

(a) Sketch the structures of the boron hydrides B_5H_9 (which has a square-based pyramidal structure, with a terminal hydrogen on each boron atom, and bridging hydrogen atoms between the four basal boron atoms) and $[B_6H_6]^{2-}$ (which has an octahedral structure with a terminal hydrogen attached to each boron atom).

(b) Demonstrate that these molecules must be **electron deficient** by counting valence electrons.

2.9 Triatomic molecules showing σ and π bonding

As with the examples in section 2.8, we have to consider the formation of **bonding**, **antibonding**, and **non-bonding** MOs. However, we must also now consider how *p* orbitals in different orientations will overlap. In exactly the same manner as we have seen for diatomic molecules (see section 2.3) p_z atomic orbitals overlap to form σ MOs while the p_x and p_y orbitals will overlap to give π MOs. Thus, in total, a triatomic molecule with both σ and π bonding will show:

two bonding σ MOs formed from *s* and p_z atomic orbitals;

two bonding π MOs formed from p_x and p_y atomic orbitals;

two non-bonding σ MOs formed from *s* and p_z atomic orbitals;

two non-bonding π MOs formed from p_x and p_y atomic orbitals;

two antibonding σ MOs formed from *s* and p_z atomic orbitals;

two antibonding π MOs formed from p_x and p_y atomic orbitals.

Worked example 2.9A

(a) Draw diagrams to show how the 2*s* and 2*p* orbitals can overlap to give bonding, non-bonding, and antibonding sigma and pi orbitals in the molecule CO_2.

(b) From your answer to part (a), count how many bonding, non-bonding, and antibonding orbitals are produced and also count how many valence electrons there are in a molecule of CO_2.

(c) Hence suggest what would happen to the bond strengths of the C═O bonds in CO_2 if the molecule were ionized to form CO_2^{+}.

Solution

(a) Let us first consider the non-bonding interactions. These arise when the orbitals on the oxygen atoms are out of phase with one another. This means that even if the orbital on the central carbon atom can interact constructively with one of the oxygen orbitals, it will interact destructively with the other, leading to no net bonding. The resultant molecular orbitals are localized on the oxygen atoms, and contain the lone pairs of electrons.

The oxygen 2*s* orbitals interact out of phase to give a non-bonding orbital:

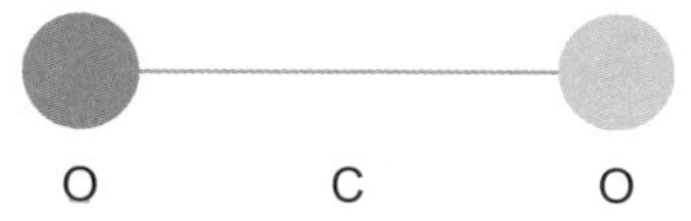

Similarly, the oxygen $2p_x$ and $2p_y$ orbitals may interact out of phase to produce a pair of degenerate non-bonding orbitals:

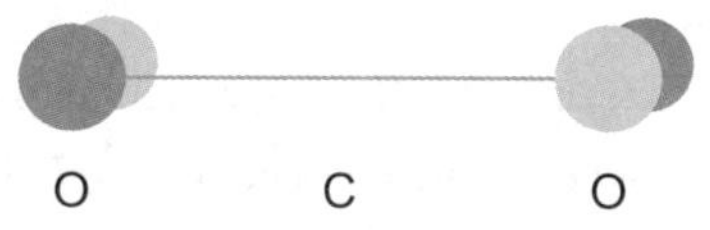

Finally, the $2p_z$ orbitals can also interact to give a non-bonding orbital:

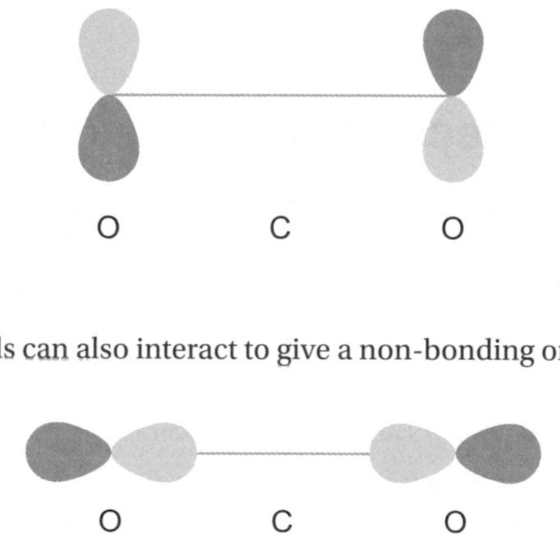

It may look as though they are interacting in phase, but remember, the non-bonding nature of this orbital is due to the fact that regardless of the phase of the central carbon $2p_z$ orbital, no net bonding interaction will be observed. As it is non-bonding, it will have the

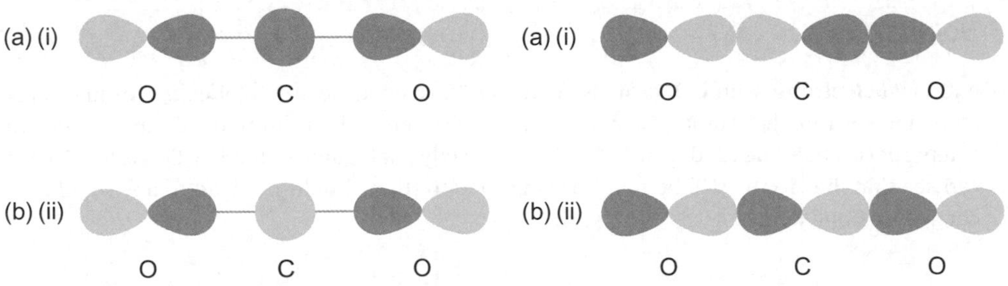

Figure 2.13 (a) interaction of the O($2p_z$) and C($2s$) orbitals giving rise to (i) bonding and (ii) antibonding σ molecular orbitals in CO_2. (b) interaction of the O($2p_z$) and C($2p_z$) orbitals giving rise to (i) bonding and (ii) antibonding σ molecular orbitals in CO_2.

Note that the $2p_z$ orbitals on O and not the $2s$ orbitals form sigma bonds to the central C atom. This is because the $2s$ orbitals are too low in energy to interact and they remain non-bonding.

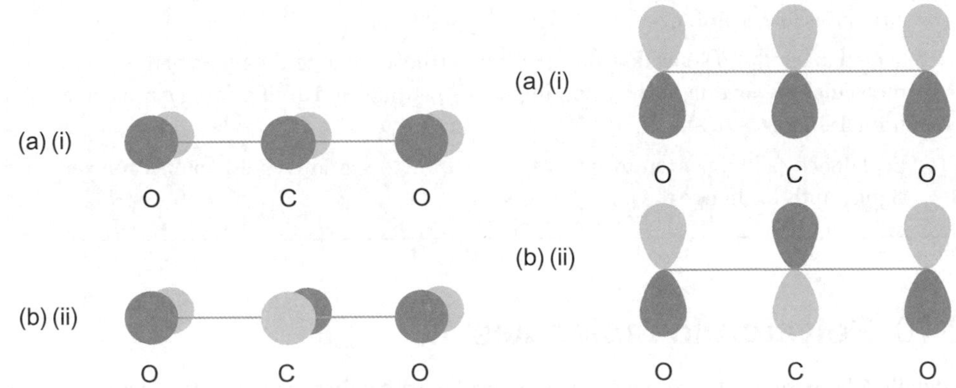

Figure 2.14 (a) interaction of the O($2p_x$) and C($2p_x$) orbitals giving rise to (i) bonding and (ii) antibonding π molecular orbitals in CO_2. (b) interaction of the O($2p_y$) and C($2p_y$) orbitals giving rise to (i) bonding and (ii) antibonding π molecular orbitals in CO_2.

same energy as the O($2p$) orbitals and hence will be degenerate with the other $2p$-derived non-bonding orbitals described previously.

To form both bonding and antibonding orbitals, the oxygen orbitals must be in phase. The nature of the resultant molecular orbital depends on whether the interaction with the carbon atom atomic orbital is constructive or destructive:

The oxygen $2p_z$ orbitals may interact with the carbon $2s$ orbital, to form bonding and antibonding σ orbitals, as illustrated in Figure 2.13a.

The oxygen $2p_z$ orbitals may interact with the carbon $2p_z$ orbital, to form bonding and antibonding σ orbitals as shown in Figure 2.13b.

The oxygen $2p_x$ and $2p_y$ orbitals may interact with the carbon $2p_x$ and $2p_y$ orbitals, to form degenerate pairs of bonding and antibonding π orbitals as shown in Figure 2.14.

(b) In total there are four bonding, four non-bonding, and four antibonding orbitals. In total, there are 16 valence electrons (four from the C atom and six from each of the O atoms). Thus we can see that all of the bonding and non-bonding orbitals are filled, leaving the antibonding orbitals vacant.

(c) The HOMO (highest occupied molecular orbital) in this molecule is non-bonding so we would not expect any significant changes in bond length upon ionization.

Worked example 2.9B

Consider the molecule N_2O. From what you have learned from Worked example 2.9A regarding the bonding in CO_2, propose a model that accounts for the π-bonding in N_2O. Do you think that the bond lengths in N_2O would change significantly if the molecule were ionized to form $[N_2O]^+$?

Solution

N_2O is **isoelectronic** with CO_2 as it also contains 16 valence electrons (N_2O: five from each N atom and six from the O atom; CO_2: six from each O atom and four from the C atom). Thus we can propose a bonding model for N_2O which is exactly analogous to that in CO_2. As for CO_2 we can see that the HOMO will be non-bonding so ionization should not lead to any significant changes in bond length.

Resonance structures are forms of a molecule where the chemical connectivity is the same but the electrons are distributed differently around the structure.

Question 2.11

(a) Sketch a diagram of the allyl anion $[C_3H_5]^-$ showing possible Lewis (resonance) structures for the anion.

(b) Sketch diagrams to show how the $2p_x$ orbitals (those at 90° to the plane of the molecule) may overlap to give bonding, non-bonding, and antibonding π molecular orbitals in this radical.

(c) Explain why adding or removing an electron from this anion would not be expected significantly to change the bond lengths.

2.10 Polyatomic molecules

A detailed discussion of the application of molecular orbital theory to polyatomic molecules is beyond the scope of this book. However, it is useful to compare two other approaches that have been used, using the penta-atomic tetrahedral molecule, methane, as an example.

Hybridization

Hybrid atomic orbitals are a useful way of describing the geometry of a range of molecules. For example, we can explain the geometry around a carbon atom in an alkane (tetrahedral), an alkene (trigonal planar), and an alkyne (linear) by making use of different **hybrid orbitals**. Figure 2.15 demonstrates how the 2*s* and 2*p* orbitals in a carbon atom can **hybridize** to form sp^3, sp^2, and *sp* orbitals respectively.

These hybrid orbitals can then be used to construct molecular orbitals following the same techniques we have used previously with standard atomic orbitals. Hybrid orbitals in this scheme have the correct symmetry to form sigma bonds, whilst any remaining non-hybridized *p* orbitals can form pi bonds as normal.

Worked example 2.10A

Consider the molecule methane. What hybridization scheme of orbitals on the central atom is implied by the tetrahedral geometry of the molecule? How would this hybridization scheme be modified to account for molecules with (a) trigonal planar and (b) linear coordination about a central carbon atom?

Solution

Referring to the diagrams in Figure 2.15 it may be seen that a molecule with tetrahedral geometry implies sp^3 hybridization on the central atom. This would be modified to sp^2 (with one unhybridized *p* orbital left over) for trigonal planar and *sp* (with two unhybridized *p* orbitals left over) for linear geometry respectively.

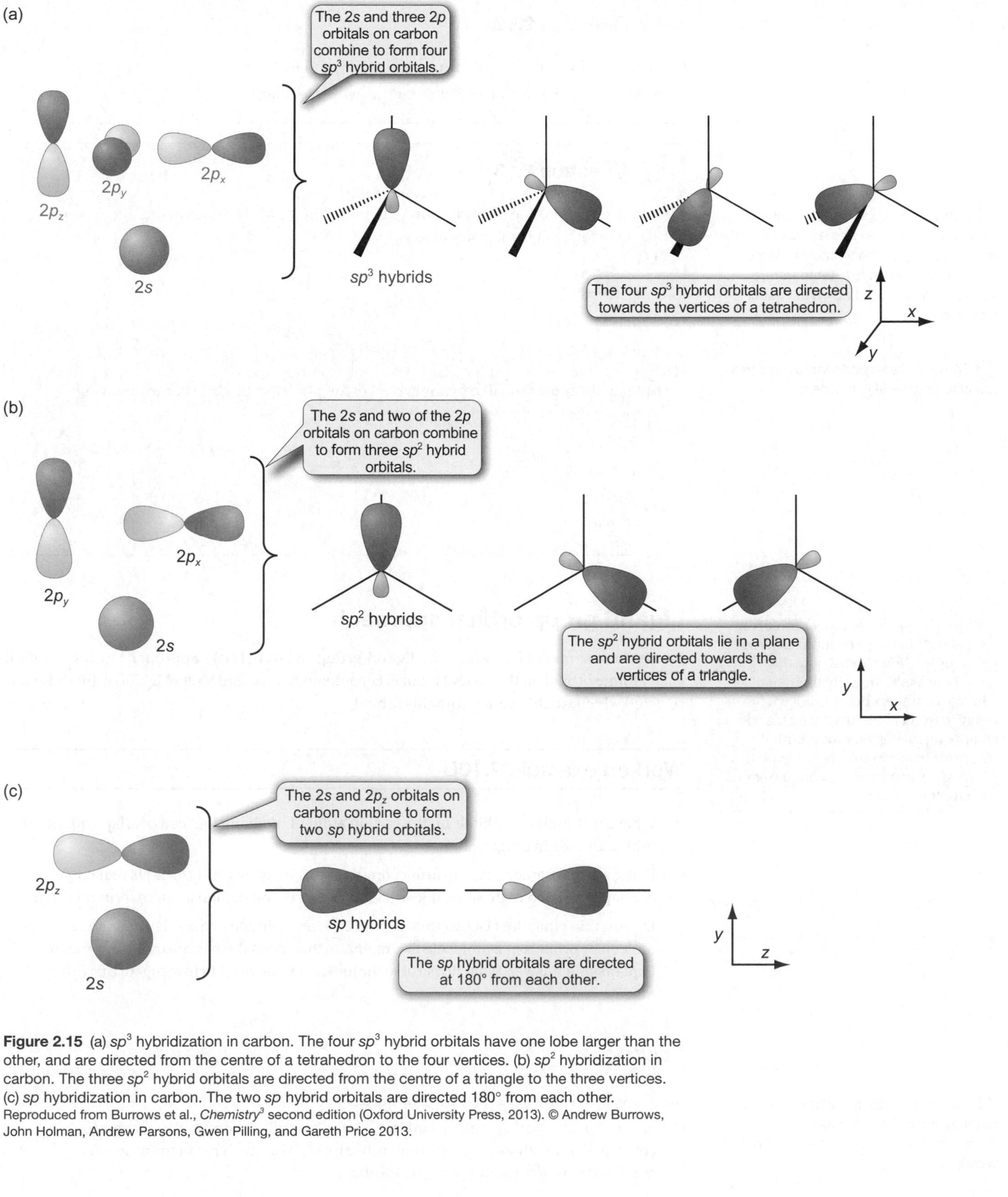

Figure 2.15 (a) sp^3 hybridization in carbon. The four sp^3 hybrid orbitals have one lobe larger than the other, and are directed from the centre of a tetrahedron to the four vertices. (b) sp^2 hybridization in carbon. The three sp^2 hybrid orbitals are directed from the centre of a triangle to the three vertices. (c) *sp* hybridization in carbon. The two *sp* hybrid orbitals are directed 180° from each other.
Reproduced from Burrows et al., *Chemistry*^{3} second edition (Oxford University Press, 2013). © Andrew Burrows, John Holman, Andrew Parsons, Gwen Pilling, and Gareth Price 2013.

Question 2.12

(a) Sketch the structures of the molecules NH_3 and BCl_3.

(b) Suggest hybridization schemes for the two molecules.

Some of these silicon species are very weakly bonded and are known only as transient gas-phase intermediates or when trapped in low-temperature solids—a technique known as matrix isolation.

Some of these phosphorus species are also very weakly bonded.

Question 2.13

(a) Suggest a sensible hybridization scheme for the *s* and *p* orbitals of the silicon atom in the following molecules or ions:

i. $SiCl_4$

ii. $O{=}SiCl_2$

iii. $O{=}Si{=}O$

iv. Si_2Cl_6

(b) State the most probable hybridization of the phosphorus atom in the following molecules or ions:

i. PCl_3

ii. $O{=}PCl$

iii. $[PCl_4]^+$

iv. PN

A ligand is an atom, ion, or molecule that binds to a central atom to form a complex. You should note that methane isn't normally thought of as being a complex but this is a useful way to think about the molecule when applying the ligand group orbital approach. For examples of ligands and of transition metal complexes see Chapter 5.

Ligand group orbital approach

An alternative method is to use the **ligand group orbital (LGO) approach**. In this method each atomic orbital on the central atom is considered in turn and a set of ligand orbitals is constructed which would overlap with this orbital.

Worked example 2.10B

(a) What are the atomic orbitals on the carbon atom of methane that can overlap with the 1*s* orbitals on the hydrogen atoms?

(b) Using a cubic framework, construct combinations of hydrogen 1*s* orbitals that could overlap with **each** of these atomic orbitals on the central carbon atom to form methane.

(c) Do you think that the LGO approach gives a better estimate of the relative energies of the molecular orbitals in a molecule like methane than does the hybridization approach? Explain your answer. What physical technique could be used to investigate this order of MOs?

Solution

The formal construction of the LGOs requires careful application of group theory, which is outside the scope of this workbook.

(a) These are clearly the 2*s* and $2p_{x,y,z}$ orbitals.

(b) The 2*s*, $2p_x$, $2p_y$, and $2p_z$ valence orbitals on the central carbon atom can combine constructively with four ligand group orbitals (LGOs), which result from various combinations of the four hydrogen 1*s* orbitals.

Symmetry considerations allow us to determine which LGOs will interact with the carbon valence orbitals to form molecular orbitals.

As can be seen in Figure 2.16, the 2*s* orbital, which is spherically symmetric, overlaps constructively with LGO(1). The $2p_x$, $2p_y$, and $2p_z$ orbitals interact constructively with LGO(2), LGO(3), and LGO(4) respectively.

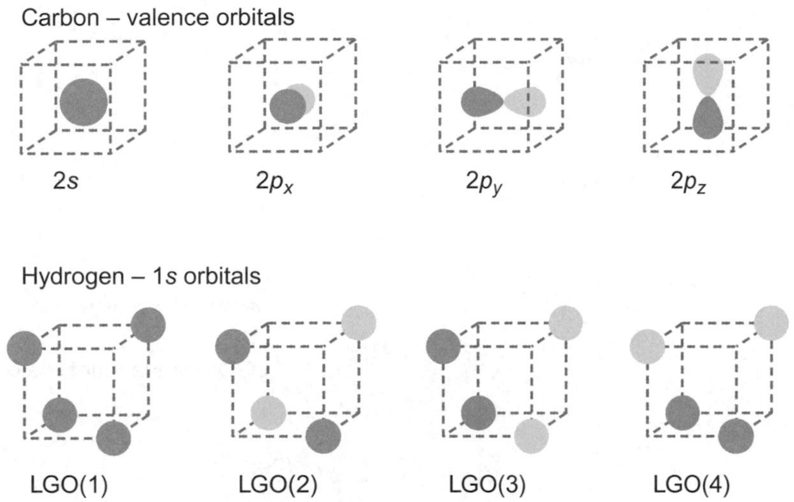

Figure 2.16 Schematic to show the carbon valence orbitals and LGOs formed from four H(1s) orbitals.

(c) To answer this, we must first ascertain how the energies of the molecular orbitals we generated using this approach compare to those generated via the hybrid orbital approach.

The overlap of the carbon 2*p* orbitals with LGO(2), LGO(3), and LGO(4) produces a set of degenerate orbitals (σ(2*p*)) that are of slightly higher energy than that generated by the overlap of 2*s* with LGO(1), σ(2*s*); see Figure 2.17. This is because the carbon 2*s* orbital is closer in energy to the hydrogen 1*s* orbitals that contribute to the LGOs, and so results in more efficient overlap leading to greater stabilization for the bonding MO.

This contrasts to the hybrid orbital approach, which results in four degenerate sp^3 hybrid orbitals on the central C atom, which then overlap with the 1*s* orbitals to produce four degenerate molecular σ bonding orbitals, as shown in Figure 2.18.

The LGO approach to constructing our MO diagram gives a better estimate of the molecular orbital energy levels. If we consider the atomic orbitals on carbon to be sp^3 hybridized then this would imply that all the molecular orbitals formed by overlap with the H 1*s* orbitals would have the same energy. However, in the LGO approach we can see that the orbital formed by overlap of C 2*s* will have a different (lower) energy than those formed by overlap of the C $2p_{x,y,z}$ orbitals. Experimental studies, using UV photoelectron spectroscopy (UPS) which determines the amount of energy required to ionize the bonding electron pairs in a molecule, have shown this to be the case, i.e. that one bonding pair of electrons in methane has a lower energy than the other three pairs. A summary of this analytical technique may be found in Box 4.8 of Burrows et al. (2013).

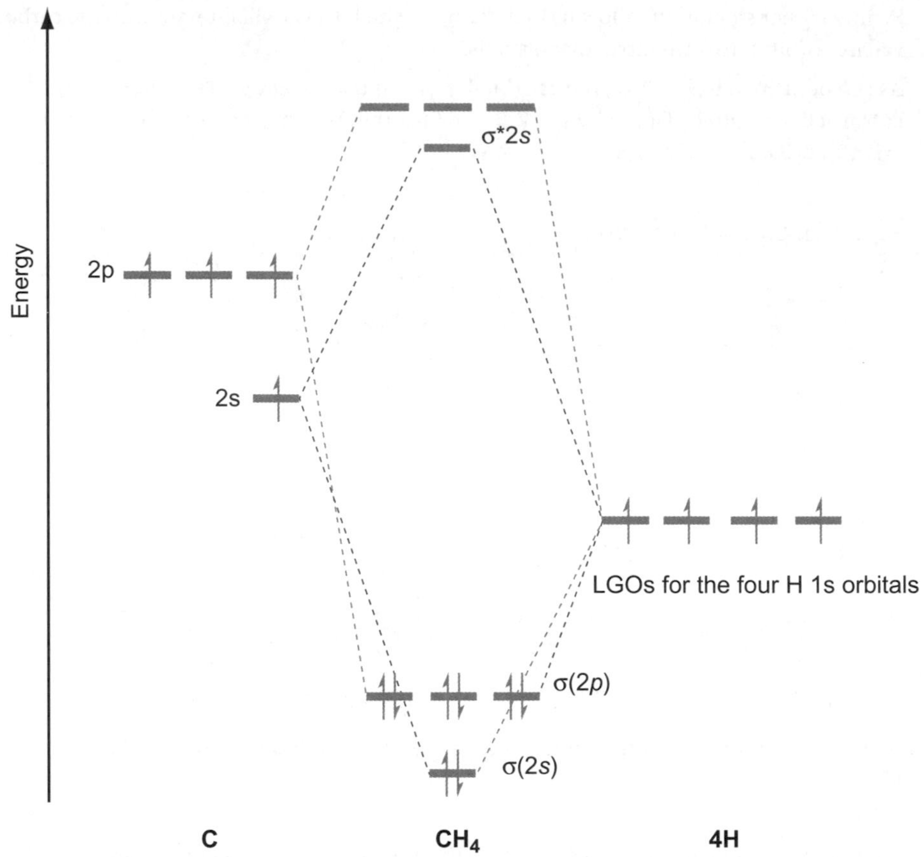

Figure 2.17 Molecular orbital energy level diagram for CH_4 produced by the LGO method.

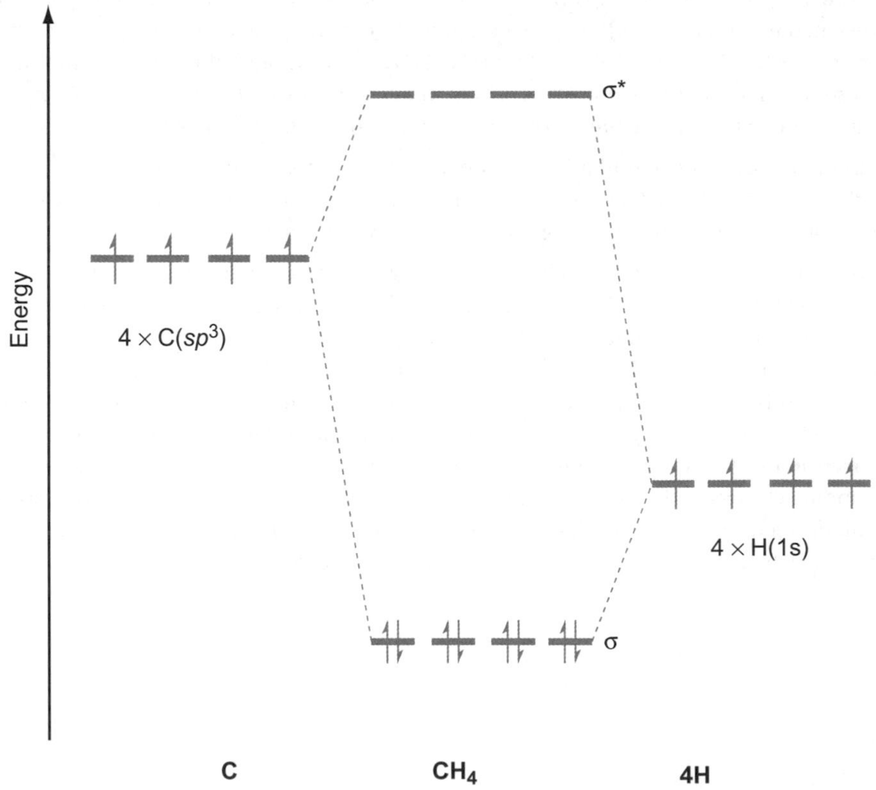

Figure 2.18 Molecular orbital energy level diagram for CH_4 produced by the hybrid orbital method.

Question 2.14

(a) Use i. hybridization and ii. the ligand group orbital approach to propose a molecular orbital diagram for SiH_4.

(b) Which of these approaches gives a better estimate of the **energies** of the bonding molecular orbital in SiH_4?

2.11 Shapes of *p*-block covalent compounds

Valence shell electron pair repulsion (VSEPR) theory

Molecular orbital theory quickly becomes complicated once we move beyond simple diatomic molecules. One useful method for determining the shapes of many *p*-block molecules and ions is known as valence shell electron pair repulsion (VSEPR) theory. Following simple rules to count the number of pairs of electrons **around a central atom** allows us to predict the shape of simple molecules.

One popular method for approaching VSEPR is the so-called **AXE method**, which is summarized briefly below:

i. Calculate the number of **valence** electrons on the **neutral** central atom (A).

ii. Determine the number of electrons associated with each ligand (X) based upon its valency (i.e. how many bonds it will form). For example, a covalently bonded F atom will form one covalent bond, to which it contributes one electron. In this case, we add one electron to the total count. An oxygen atom typically forms a double bond and hence contributes two electrons to the total count. In that case, we add two electrons to the total count.

iii. Subtract electrons to account for a positive charge (i.e. for a +2 charge, subtract two electrons) or add electrons to account for a negative charge (i.e. add two electrons for a –2 charge).

iv. Subtract $2(n-1)$ electrons for each *n*th order bond (i.e. for a triple bond, subtract $2 \times (3-1) = 4$ electrons).

v. Calculate the total number of electron pairs around the central atom by dividing the total number of electrons calculated in steps i to iv by two. The total number of **pairs** of electrons is called the **steric number** which allows us to get the base shape for the molecule.

vi. Determine how these electrons are distributed between bonding pairs (X) and lone pairs (E). The number of lone pairs equals the total number of electron pairs minus the number of bonded atoms.

vii. Place the pairs of electrons around the central atom according to the base shape, keeping in mind that electronic repulsion decreases according to:

$$E-E > E-X > X-X.$$

viii. The final shape of the molecule **does not** include the lone pairs.

Table 2.1 provides a series of commonly found base shapes, their associated steric number, and their 'AXE' configuration.

Table 2.1 Commonly found base shapes, their associated steric number and their 'AXE' configuration.

Steric number	0 lone pairs (Base shape)	1 lone pair	2 lone pairs	3 lone pairs
2	 AX_2E_0 Linear			
3	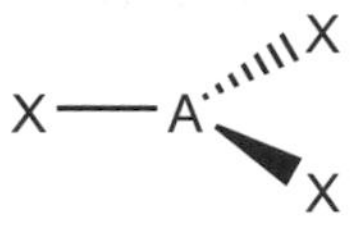 AX_3E_0 Trigonal planar	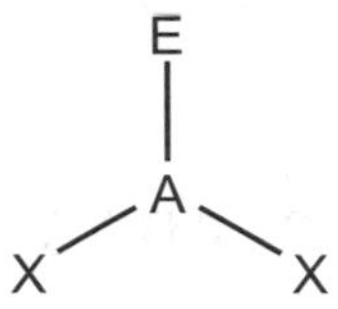 AX_2E_1 Bent		
4	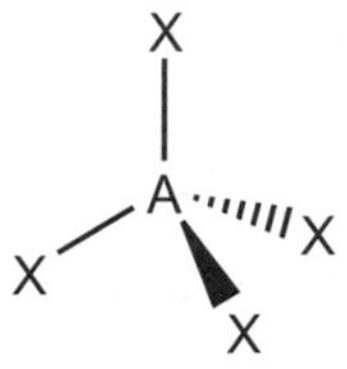 AX_4E_0 Tetrahedral	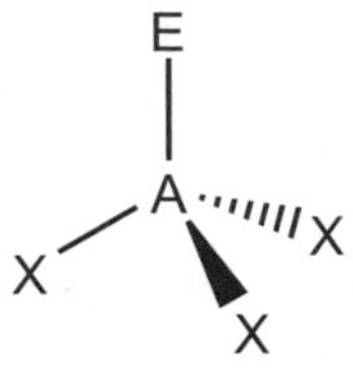 AX_3E_1 Trigonal pyramidal	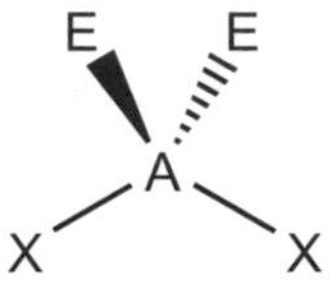 AX_2E_2 Bent	
5	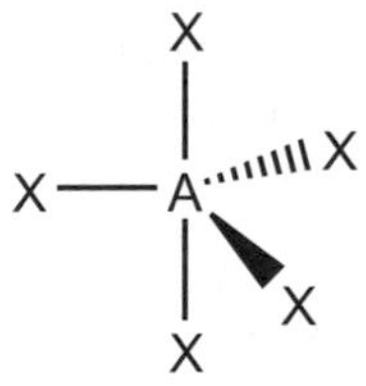 AX_5E_0 Pentagonal bipyramidal	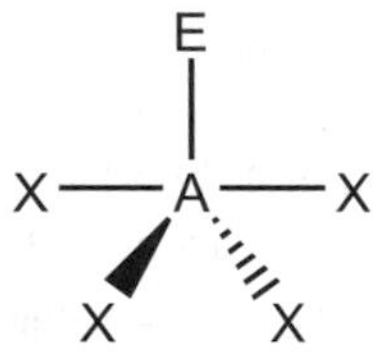 AX_4E_1 See-saw	 AX_3E_2 T-shaped	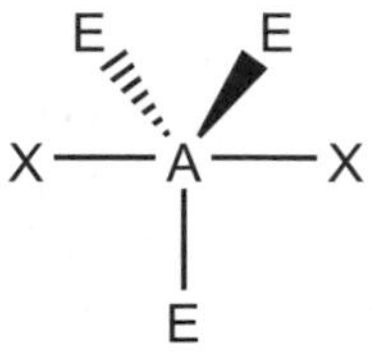 AX_2E_3 Linear
6	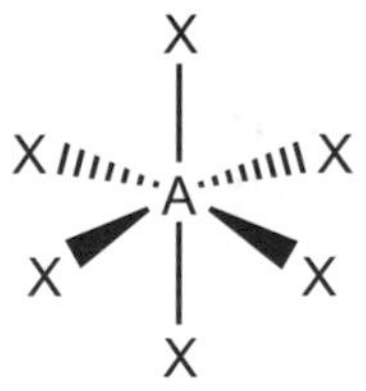 AX_6E_0 Octahedral	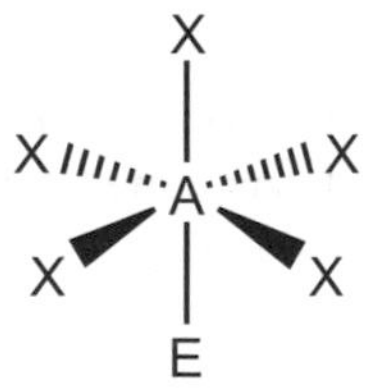 AX_5E_1 Square pyramidal	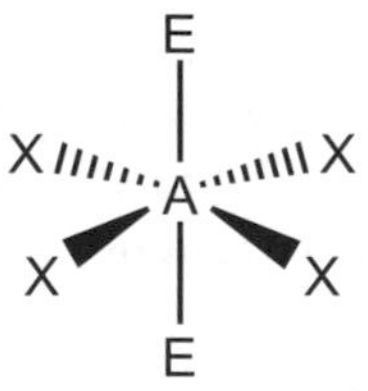 AX_4E_2 Square planar	
7	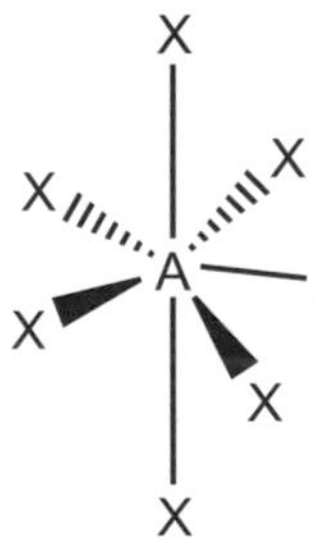 AX_7E_0 Pentagonal bipyramidal			

Worked example 2.11A

(a) Use valence shell electron pair repulsion (VSEPR) theory to predict the shapes of the following molecules. Sketch the structure of each molecule and briefly explain how VSEPR accounts for your proposed structure.

i. IF_5

ii. BrF_3

iii. XeF_4

iv. SF_4

(b) Use VSEPR theory to propose the shapes adopted by the ions IF_4^+ and I_3^+. State clearly how you determine the structures that you propose.

(c) Use VSEPR theory to predict the shapes of $[GaCl_4]^-$ and $[GaCl_5]^{2-}$.

(d) Use VSEPR theory to propose the shape adopted by the iodate ion IO_3^-. State clearly how you determine the structure that you propose. Discuss which atomic orbitals on the iodine atom may be used to form multiple bonds in IO_3^-.

(e) Would you expect the bond angles in IF_5 and XeF_4 respectively to be exactly the same as the bond angles in a regular octahedron (90°) or somewhat larger or smaller than 90°?

Solution

(a) In the first part of this question some simple molecules with single bonds to fluorine are considered. First we must count the electrons on the central atom. Then we must determine how many of these electrons form bonds and how may are non-bonding lone pairs. We may then determine how these electron pairs will be distributed around the central atom. Ignoring the lone pair then gives us the geometry of the atoms around the central atom, i.e. the shape of the molecule.

Using the AXE method:

IF_5	Number of electrons
Valence electrons on I	7
Electrons from each F	$1 \times 5 = 5$
0 charge	0
0 *n*th order bonds	0
Total number of electrons	12

In IF_5 we have seven electrons on the central I atom. Five of these electrons form covalent bonds to F, which each contribute one electron to the total count. This gives us a total of 12 electrons and therefore six pairs of electrons, so our molecule will be based on an octahedral distribution of electrons. We have five bonding pairs of electrons (one for each of the fluorine atoms) and hence one lone pair of electrons. This gives us the AX_5E_1 configuration; ignoring the lone pair of electrons gives a square pyramidal structure.

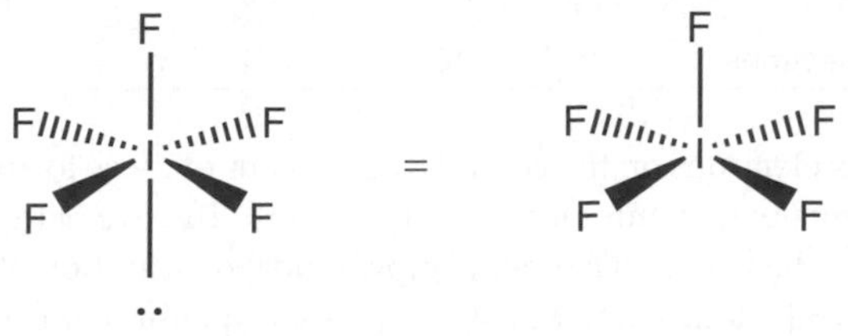

BrF_3	**Number of electrons**
Valence electrons on Br	7
Electrons from each F	$1 \times 3 = 3$
0 charge	0
0 *n*th order bonds	0
Total number of electrons	10

In BrF_3 we again have seven electrons on the central atom (Br), and three of these form covalent bonds to fluorine. This gives 10 electrons in total and therefore five pairs of electrons. Thus the structure is based on a trigonal bipyramidal distribution of electrons. The AX_3E_2 configuration means that lone pairs occupy *cis* equatorial positions so the overall structure is 'T'-shaped.

XeF_4	**Number of electrons**
Valence electrons on Xe	8
Electrons from each F	$1 \times 4 = 4$
0 charge	0
0 *n*th order bonds	0
Total number of electrons	12

In XeF_4 there are eight electrons on the central atom. Four of these form covalent bonds with fluorine, increasing the total number of electrons to 12. The six pairs of electrons mean that the structure is based on an octahedral distribution of electrons. Given that the molecule has two lone pairs (AX_4E_2), the lone pairs occupy *trans* positions so the overall structure is square planar.

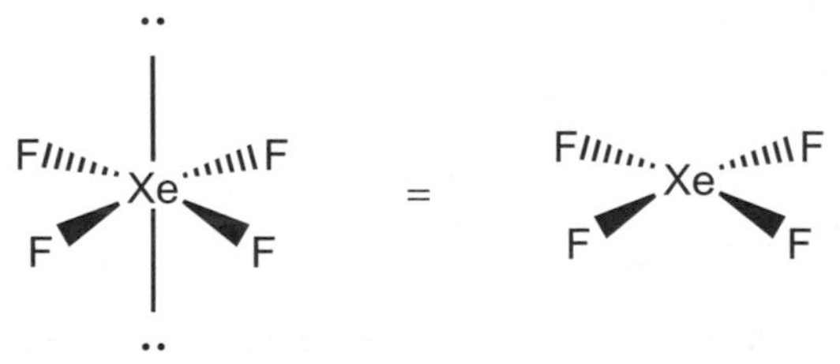

SF_4	**Number of electrons**
Valence electrons on S	6
Electrons from each F	$1 \times 4 = 4$
0 charge	0
0 *n*th order bonds	0
Total number of electrons	10

In SF_4 there are six electrons on the central S atom. Four of these form covalent bonds to fluorine, increasing the total number of electrons to 10. The five pairs of electrons mean that the structure is based upon a trigonal bipyramidal distribution of electrons around the central S atom. The lone pair (AX_4E_1) occupies an equatorial position and the structure is 'see-saw'.

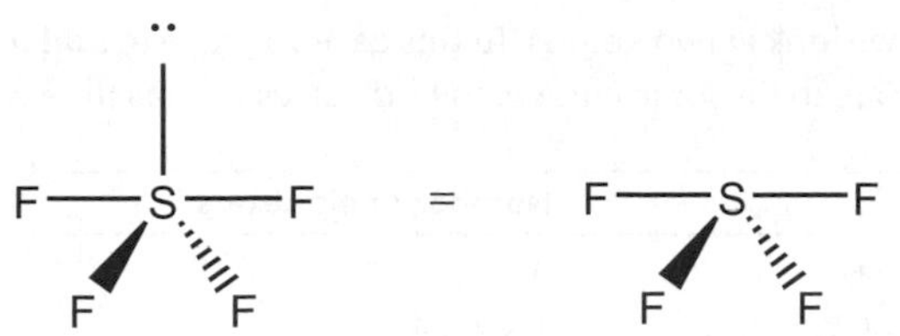

(b) In this part of the question, we now must consider the effect of having a positive ionic charge. In order to accommodate this we simply have to **remove** the appropriate number of electrons corresponding to the charge on the ion. A $+n$ charge means we must remove n electrons.

IF_4^+	**Number of electrons**
Valence electrons on I	7
Electrons from each F	$1 \times 4 = 4$
+1 charge	−1
0 *n*th order bonds	0
Total number of electrons	10

In IF_4^+ there are seven electrons on the central atom. The covalent bonds formed with the four fluorine atoms results in 11 electrons in total. To account for the +1 charge, we subtract one electron, to give 10 electrons in total. There are therefore four bonding pairs and one lone pair. The AX_4E_1 structure is therefore 'see-saw'—exactly analogous to SF_4. It may be noted that IF_4^+ and SF_4 are **isoelectronic**.

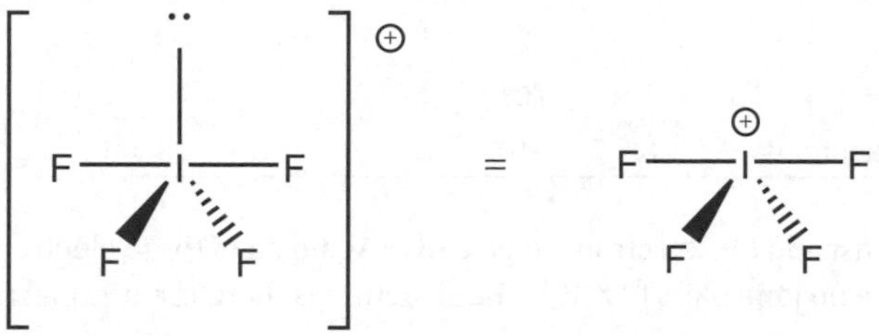

I_3^+	**Number of electrons**
Valence electrons on I	7
Electrons from each I	$1 \times 2 = 2$
+1 charge	−1
0 *n*th order bonds	0
Total number of electrons	8

In I_3^+ there are seven electrons on the central atom, plus an additional two for each covalently bonded iodine atom. Subtracting one for the charge leaves four pairs of electrons in total and therefore this has two lone pairs. Thus the structure is based on a tetrahedral arrangement of the four electron pairs (AX_2E_2) and hence the molecule is bent.

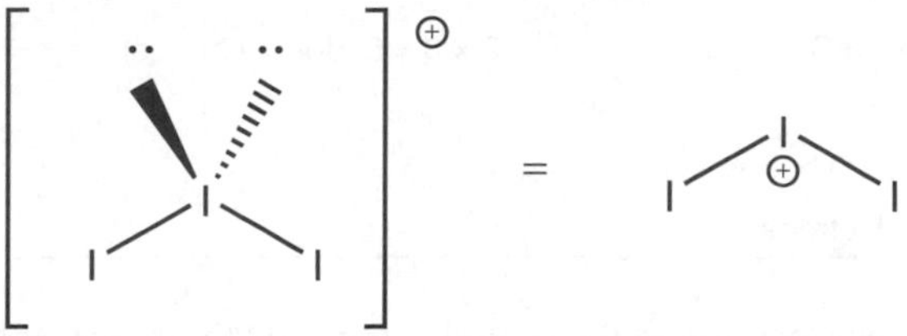

(c) In the next example we look at two anions. In this case we have to **add** to the total number of electrons appropriate to the ionic charge. In both cases Ga has three valence electrons.

$[GaCl_4]^-$	Number of electrons
Valence electrons on Ga	3
Electrons from each Cl	$1 \times 4 = 4$
–1 charge	1
0 *n*th order bonds	0
Total number of electrons	8

In $[GaCl_4]^-$ we add one electron to give four. All of these form bonds to Cl, so there are no lone pairs (AX_4E_0) and therefore the structure is tetrahedral.

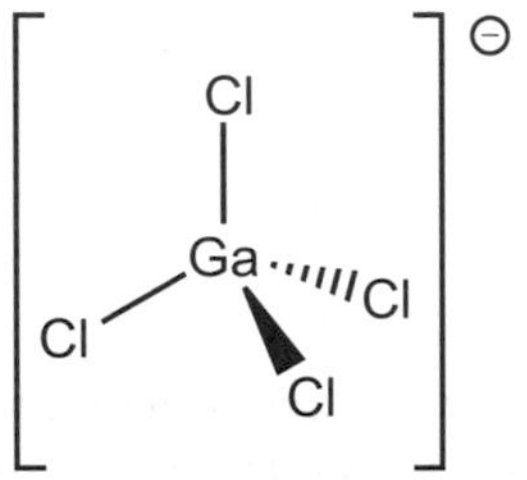

$[GaCl_5]^{2-}$	Number of electrons
Valence electrons on Ga	3
Electrons from each Cl	$1 \times 5 = 5$
–2 charge	2
0 *n*th order bonds	0
Total number of electrons	10

In $[GaCl_5]^{2-}$ we must add two electrons to give five. Again all of these electrons form bonds to Cl and there are no lone pairs (AX_5E_0). The structure is therefore trigonal bipyramidal.

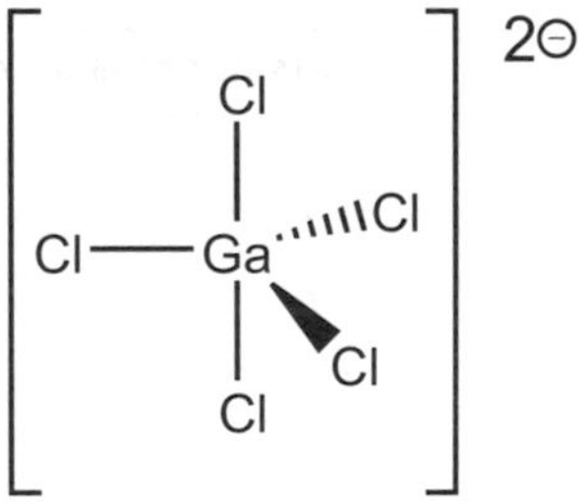

(d) In the final example we consider a case where not only is there an ionic charge but also there is multiple (in this case double) bonding.

IO_3^-	Number of electrons
Valence electrons on I	7
Electrons from each O	$2 \times 3 = 6$ (double bonds)
–1 charge	1
Three 2nd order I=O bonds	-2×3
Total number of electrons	8

Here we have seven electrons on the central I atom, and three covalently bonded oxygen atoms. If we assume that the negative charge resides on the central I atom, then each oxygen forms a double bond to the iodine. This gives 13 electrons, and adding one electron for the negative charge gives fourteen. Given that we have double bonds, we must remove electrons to account for this. Subtracting six electrons for the three second-order I=O bonds leaves a total of eight electrons. There are three bonding pairs and one lone pair, which gives a structure based on a tetrahedron. The AX_3E_1 configuration means that the overall structure of the ion is trigonal pyramidal.

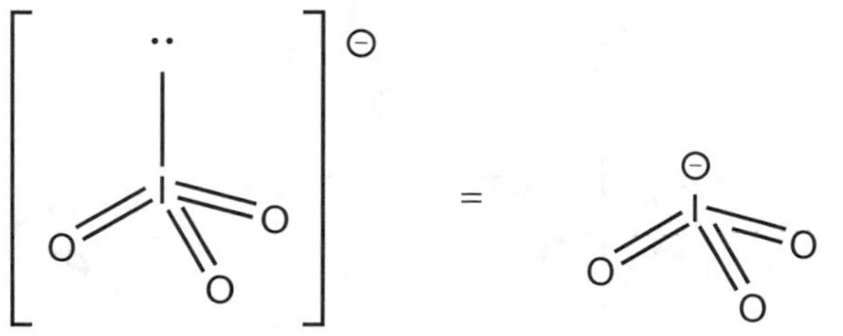

We can also approach this from a different perspective. Knowing that in reality, the negative charge is more reasonably assigned to reside on one of the oxygen atoms (due to the higher electronegativity of oxygen than iodine) we would use the following table:

IO_3^-	Number of electrons
Valence electrons on I	7
Electrons from each O	$2 \times 2 = 4$ (double bonds) $1 \times 1 = 1$ (single bond)
0 charge on I	0
Two 2nd order I=O bonds	-2×2
Total number of electrons	8

As can be seen, we end up with the same **steric number** and the same AX_3E_1 configuration that we identified previously, and hence obtain the same trigonal pyramidal shape for the ion:

In reality, the negative charge is delocalized over all of the oxygen atoms through resonance effects, and hence when measured experimentally, all bond lengths are seen to be equivalent.

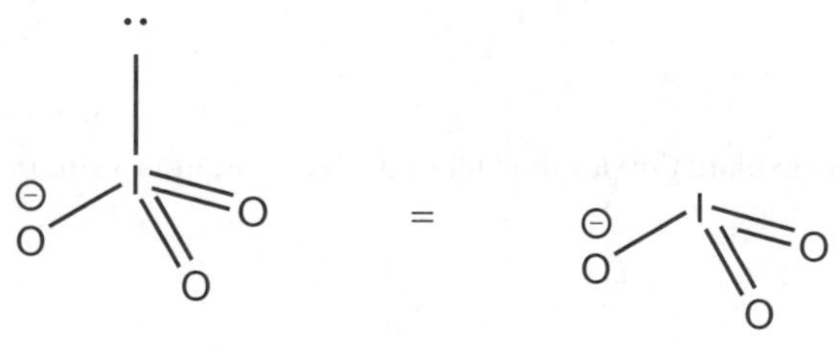

(e) In part (a) we determined that the steric number for IF_5 and XeF_4 is six, i.e. there are six pairs of electrons to distribute around the central atom and hence these structures have a shape based on an octahedron.

When considering the bond angles using VSEPR, it is important to remember that lone pairs of electrons (E) repel other pairs of electrons more strongly than bonding pairs (X) as they are held more closely to the central atom. The strength of repulsion is therefore ranked as follows: $E-E > E-X > X-X$.

This rule not only helps us to determine where best to place the lone pairs and bonding pairs around the central atom, but also allows us to make predictions about bond angles.

For IF_5 (Figure 2.19a) it can be seen that the axial lone pair will repel the equatorial F atoms more strongly than the axial F atom. This means that the bond angle between the axial fluorine atom and equatorial fluorine atoms should be expected to be less than 90°, whilst the angle between neighbouring equatorial fluorine atoms should be expected to remain 90°.

The terms *axial* and *equatorial* are used to describe the positions of atoms within a molecule. In IF_5 the equatorial F atoms lie within the xy plane while the two axial F atoms lie along the z axis.

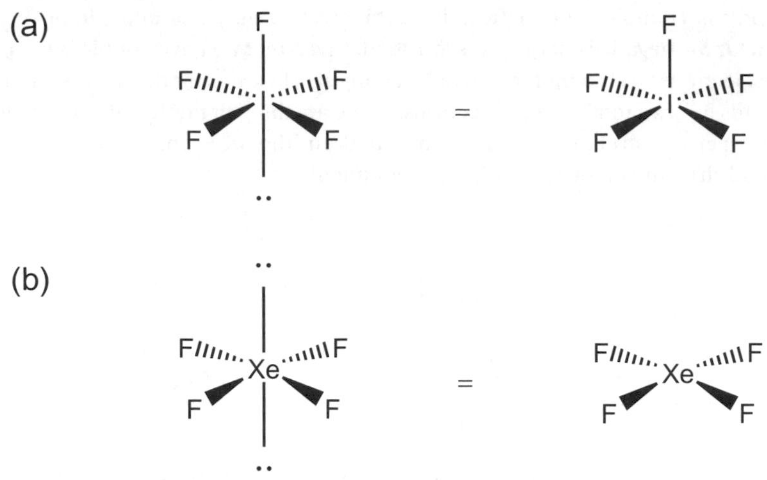

Figure 2.19 (a) The shape of IF_5 as predicted by VSEPR is square-based pyramidal. (b) The shape of XeF_4 as predicted by VSEPR is square planar.

For XeF_4 (Figure 2.19b), both lone pairs are axial, and repel the equatorial F atoms equally. As such, the FXeF bond angles should be expected to remain 90°.

Worked example 2.11B

(a) Use VESPR theory to suggest structures for the following molecules and ions:

i. $[XeO_6]^{4-}$
ii. $XeOF_2$
iii. $XeOF_4$
iv. XeO_2F_2
v. XeO_2F_4
vi. XeO_3F_2

(b) Suggest a possible explanation for the fact that $[XeF_8]^{2-}$ adopts a square antiprismatic structure.

Solution

(a) All of these examples are taken from xenon chemistry. They are somewhat complex because they contain double bonds to oxygen and in two cases there is an ionic charge. It should also be noted that the number of electrons to consider in xenon compounds will typically be high as a xenon atom itself has eight valence electrons.

i. Using the AXE method, we can see that by following the general rules we have encountered previously and without taking chemical knowledge into account, we arrive at a total of 12 electrons, or six pairs of electrons distributed around the central atom. Given that there are six bonded atoms, we end up with an AX_6E_0 configuration, giving an octahedral molecule:

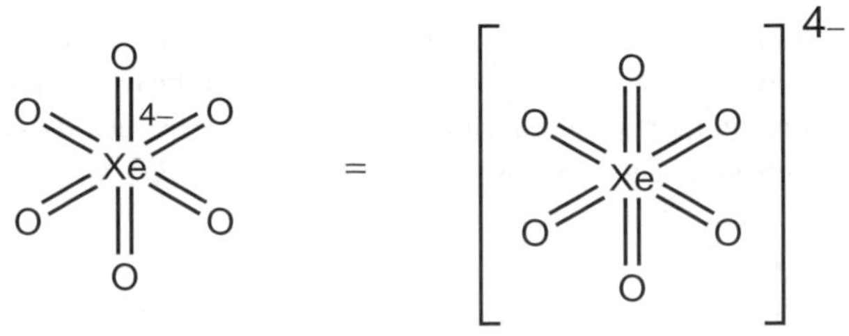

$[XeO_6]^{4-}$	Number of electrons
Valence electrons on Xe	8
Electrons from each O	2 × 6 = 12 (double bonds)
–4 charge	4
Six 2nd order Xe═O bonds	–2 × 6
Total number of electrons	12

Our chemical intuition should tell us that it is unlikely that the negative charge will sit directly on the xenon atom. Instead, a more plausible scenario can be proposed:

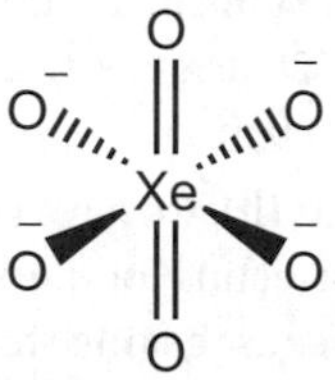

However, taking this into account still leads to the same shape, as we see in the table below. In this example, we consider the xenon atom to have retained all of its electrons (i.e. 0 charge on Xe), with the 4– charge accounted for by four of the six oxygen atoms forming single bonds. The total number of electrons obtained is the same, and hence so is the predicted geometry.

$[XeO_6]^{4-}$ – Two Xe═O bonds	Number of electrons
Valence electrons on Xe	8
Electrons from each O	2 × 2 = 4 (double bonds) 1 × 4 = 4 (single bonds)
0 charge on Xe	0
Two 2nd order Xe═O bonds	–2 × 2
Total number of electrons	12

Additional experimental information is required to determine the correct shape. The xenon–oxygen bonds in the perxenate anion have been shown to all be equivalent, and hence it seems likely that a resonance-based stabilization of the second proposed structure is likely.

ii. Again, using the AXE method for electron counting, we fill in our table using the following information: Xe has eight valence electrons, the oxygen will donate two electrons to form a double bond, whilst each of the fluorine atoms donate one electron to their respective single covalent bonds. Correcting for the presence of the double bond leads us to 10 electrons in total, or five pairs of electrons. Given that we have three atoms bonded to the central xenon, we arrive at an AX_3E_2 configuration, which means we have a T-shaped molecule:

$XeOF_2$	Number of electrons
Valence electrons on Xe	8
Electrons from O and F	$2 \times 1 = 2$ (O, double bond) $1 \times 2 = 2$ (F, single bond)
0 charge	0
One 2nd order Xe═O bond	-2×1
Total number of electrons	10

The Xe═O double bond is shorter than the Xe—F single bonds (171 pm vs 190 pm), and hence the oxygen atom is placed equatorially on the trigonal bipyramidal base shape in order to minimize repulsion with the lone pairs. This position gives an angle between the oxygen and lone pairs of about 120 degrees, whilst an axial position would place the oxygen about 90 degrees from a lone pair.

iii. Building on the method used for part (ii), we have two additional fluorine atoms bonded to the central xenon, which after correcting for the Xe═O double bond gives 12 electrons in total. The six pairs of electrons means that the shape is based on an octahedron, and given the five bonded atoms, we arrive at an AX_5E_1 configuration, which is a square-based pyramidal structure. Building on the method used for part (ii), we have an additional two fluorine atoms bonded to the central xenon, which after correcting for the Xe═O double bond gives 12 electrons in total. The six pairs of electrons mean that the shape is based on an octahedron, and given the five bonded atoms, we arrive at an AX_5E_1 configuration, which is a square-based pyramidal structure:

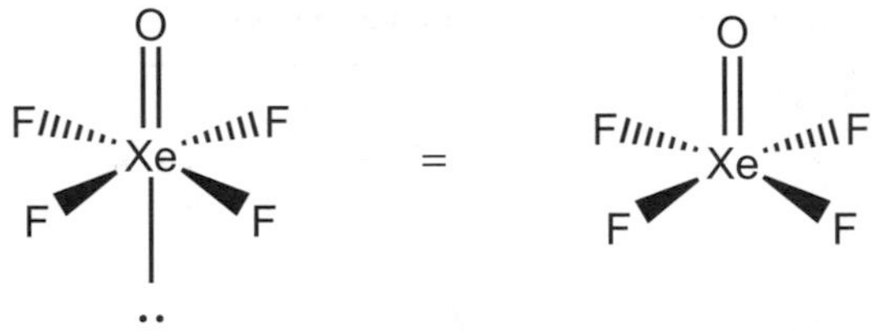

$XeOF_4$	Number of electrons
Valence electrons on Xe	8
Electrons from O and F	$2 \times 1 = 2$ (O, double bond) $1 \times 4 = 4$ (F, single bond)
0 charge	0
One 2nd order Xe═O bond	-2×1
Total number of electrons	12

The Xe═O double bond is shorter than the Xe—F single bonds (171 pm vs 190 pm), and hence the oxygen atom is placed opposite to the lone pair in order to minimize repulsion.

iv. We now have two double bonds to contend with. After counting electrons as before, we obtain an AX_4E_1 configuration, which gives a 'see-saw' shaped molecule.

XeO_2F_2	Number of electrons
Valence electrons on Xe	8
Electrons from O and F	2 × 2 = 4 (O, double bond) 1 × 2 = 2 (F, single bond)
0 charge	0
Two 2nd order Xe=O bonds	−2 × 2
Total number of electrons	10

Here, the two fluorine atoms occupy the axial positions, whilst the oxygen atoms and lone pair occupy the equatorial positions. The Xe=O double bonds (171 pm) are shorter than the Xe—F single bonds (190 pm), and hence occupy the equatorial positions to minimize the repulsion with the lone pair (bond angle at equatorial is c. 120 degrees from the lone pair, at axial it would be 90 degrees from the lone pair).

v. The electron counting scheme using the AXE method results in a total of 12 electrons or six pairs of electrons distributed around the central xenon atom. This means that the shape is based on an octahedron. There are no lone pairs, and hence the AX_6E_0 octahedral structure is observed.

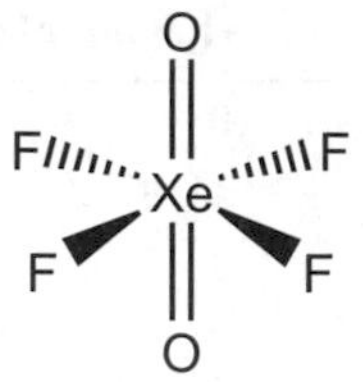

XeO_2F_4	Number of electrons
Valence electrons on Xe	8
Electrons from O and F	2 × 2 = 4 (O, double bond) 1 × 4 = 4 (F, single bond)
0 charge	0
Two 2nd order Xe=O bonds	−2 × 2
Total number of electrons	12

This structure could have *cis* or *trans* isomers where the O atoms are either adjacent to each other or opposite one another. It is interesting to think of experiments that could be used to distinguish these isomers. You may have covered IR and NMR spectroscopy. In IR spectroscopic experiments, the *cis* isomer would be expected to show two IR active Xe=O vibrations (symmetric and antisymmetric) while the *trans* isomer would have only one IR active Xe=O stretch—the antisymmetric stretch. ^{19}F NMR spectroscopy would show the number of unique F environments in the molecule. The *cis* isomer would show two F environments (fluorine atoms adjacent to the oxygens and opposite the oxygens would not be equivalent) whilst the *trans* isomer would only show one (as all fluorine atoms would be equivalent). Given that the bond length of the Xe=O double bond is less than that of the Xe—F single bond, electrostatic repulsion is expected to be minimized if the *trans* isomer is adopted.

vi. Electron counting using the AXE method leads to an AX_5E_0 configuration and hence the structure is based on a trigonal bipyramid.

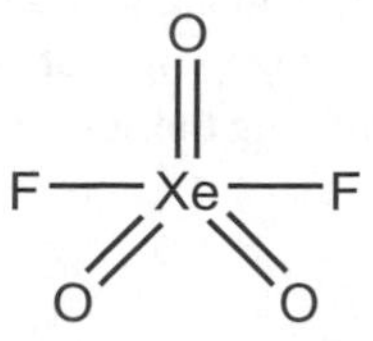

As with the answer to part (v), it is interesting to consider possible isomers and how these might be distinguished from each other.

XeO_3F_2	Number of electrons
Valence electrons on Xe	8
Electrons from O and F	2 × 3 = 6 (O, double bond) 1 × 2 = 2 (F, single bond)
0 charge	0
Three 2nd order Xe=O bonds	−2 × 3
Total number of electrons	10

Given that the bond length of the Xe=O double bond is less than that of the Xe-F single bond, electrostatic repulsion is expected to be minimized if the oxygen atoms occupy the equatorial positions in the trigonal bipyramidal structure.

(b) This is an unusual structure, with a very high steric number. Using the AXE method, we obtain a total of 18 electrons around the central xenon, or nine pairs of electrons. Given that we have eight bonded atoms, we arrive at an AX_8E_1 configuration.

$[XeF_8]^{2-}$	Number of electrons
Valence electrons on Xe	8
Electrons from F	1 × 8 = 8
−2 charge	2
0 *n*th order bonds	0
Total number of electrons	18

The most efficient way to distribute eight pairs of electrons about a central atom corresponds to a square antiprismatic structure, whilst nine pairs of electrons corresponds to a tricapped trigonal prismatic structure or a capped square antiprismatic structure (see Figure 2.20).

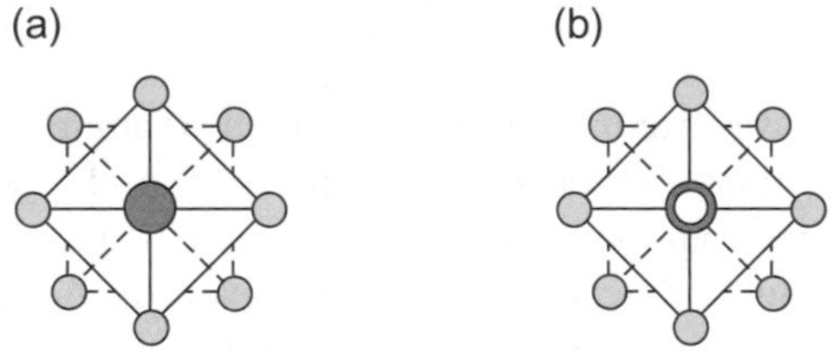

Figure 2.20 Structures predicted by VSEPR with eight pairs of electrons are based on (a) a square antiprism. Those with nine pairs of electrons are based on (b) a capped square antiprism (or the closely related tricapped trigonal prism). The figures are in top-down perspective.

This ion adopts a square antiprismatic structure as shown in Figure 2.21, so we can conclude that the lone pair is not stereochemically active. A possible explanation for this is that the lone pair is held in the 5*s* orbital and is therefore tightly held to the nucleus. This would then be an example of the 'inert pair' effect (see Chapter 3).

The atom in the centre of the structure in Figure 2.21 represents xenon, whilst surrounding atoms represent fluorine. The image to the right of the structure shows a 'top-down' perspective of the molecule, highlighting the square antiprismatic structure.

The inert pair effect is discussed in more detail in Chapter 3. This is the tendency of the outermost *s* electrons of an atom to remain unionized or unshared in the *p*-block elements. The tendency becomes more pronounced as a group of the periodic table is descended.

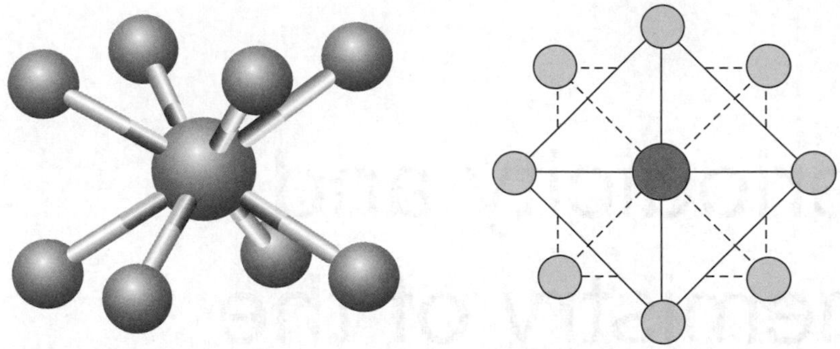

Figure 2.21 The shape of $[XeF_8]^{2-}$ as predicted by VSEPR is a tricapped trigonal prismatic structure or a capped square antiprismatic structure. However, the structure it adopts, and which is shown here, is a square antiprismatic structure.

Question 2.15

Use VSEPR to predict the structures of the following molecules and ions:

(a) H_2Te
(b) $[H_3S]^+$
(c) TeF_6
(d) TeF_4
(e) SO_3
(f) SO_2

Question 2.16

Use VSEPR to predict the structures of the following molecules and ions:

(a) $[PBr_4]^+$
(b) $[PCl_6]^-$
(c) OCS
(d) $[SnCl_5]^-$
(e) $[SiH_3]^-$
(f) $[SnS_4]^{4-}$
(g) $Si_2O_2Cl_4$

Turn to the Synoptic questions section on page 148 to attempt questions that encourage you to draw on concepts and problem-solving strategies from several topics within a given chapter to come to a final answer.

Final answers to numerical questions appear at the end of the book, and fully worked solutions appear on the book's website. Go to http://www.oxfordtextbooks.co.uk/orc/chemworkbooks/.

Reference

Burrows, A., Holman, J., Parsons, A., Pilling, G., and Price, G. (2013) *Chemistry*[3], 2nd edn (Oxford University Press, Oxford).

3

Periodicity and chemistry of the *s* and *p* blocks

3.1 **Periodicity**

The original construction of the periodic table depended to a large extent upon observation of trends in the chemistry of the elements. Subsequently, understanding of atomic theory (see Chapter 1) has allowed the theoretical background to these trends and the structure of the periodic table to be understood much more fully. The trends in chemistry on moving down a group or across a period are the basis of the concept of periodicity. Group trends are considered in the subsequent sections of this chapter. In this section the physical basis for the trends in chemistry seen on moving across a period are considered. Three fundamental properties will be considered in detail: atomic and ionic radii, ionization enthalpies, and electron affinities (electron gain energies).

➔ The terms *electron affinity and electron gain energy* refer to the process in which a gaseous atom accepts an electron to form a singly charged anion. The terms are often used interchangeably. Although electron gain energy is sometimes preferred because the sign is unambiguous, electron *affinity* will be used throughout this chapter as this term is probably more familiar to readers.

Worked example 3.1A

(a) Look up values for i. the atomic radii and ii. the first ionization enthalpies, i.e. the enthalpy for the process:

$$M(g) \rightarrow M^{+}(g) + e^{-}$$

for the elements in the second period of the periodic table (Li, Be, B, C, N, O, F, and Ne). Plot these values and comment on the overall trends and any specific anomalies that are seen in this overall trend, i.e. any points at which your graph deviates markedly from a straight line.

(b) Look up the first electron affinities, i.e. the energy for the process:

$$M(g) + e^{-} \rightarrow M^{-}(g)$$

for the elements in the second period of the periodic table (Li, Be, B, C, N, O, F, and Ne). Comment on the values in terms of the filling of atomic orbitals of these atoms by electrons.

(c) Briefly discuss these values in terms of the observation that cations are normally observed more commonly for elements at the left-hand side of the periodic table and anions for elements at the right-hand side. Covalent chemistry also tends to predominate at the right-hand side.

Solution

The relevant values to answer this question are shown in Table 3.1.

Table 3.1 Data for solution to Worked example 3.1A.

Element	Atomic radius/pm	First ionization enthalpy/ kJ mol^{-1}	First electron affinity/ kJ mol^{-1}
Li	157	520	–60
Be	112	900	>0
B	88	801	–27
C	77	1086	–122
N	74	1402	>0
O	66	1314	–141
F	64	1681	–328
Ne	154	2081	>0

(a)

i. A plot of atomic radii as we move across the second period of the periodic table is shown in Figure 3.1.

The atomic radius of an element is defined as half of the distance between neighbouring nuclei in the pure element. If the element is a metal then this distance is known as the **metallic radius**. If the element is a non-metal with covalently bonded molecules then the distance is known as the **covalent radius**. The main trend that can be seen for the elements Li to F is that as we move from left to right across the period, the atomic radius **decreases**. This is understandable in terms of the **effective nuclear charge (Z_{eff})** which has also been discussed in Chapter 1. As we move across the period the nuclear charge increases. This is not balanced by the shielding from the extra electrons which are being added to the atom and as such, the effective nuclear charge increases and the outermost electrons are pulled more closely to the nucleus; hence the atomic radius deceases.

The group 18 elements are monatomic gases and the atomic radii quoted here are **van der Waals radii**. Van der Waals radii are calculated as half the distance between the nuclei of atoms (which are not chemically bonded to each other in the case of a **monatomic** element) in the solid. As there is no bonding between these atoms the radii are considerably larger than the covalent or metallic radii quoted for the other atoms in the period.

From the graph in Figure 3.1 it appears that the atomic radius of neon is much larger than that of fluorine. However, one must be careful when looking at data for atomic radii. Two measures are commonly made. These are the *covalent radius*—a measure of the size of an atom that forms part of a covalent bond and *van der Waals radius*—the radius of an imaginary hard sphere calculated as half the distance between the nuclei of atoms which are not chemically bonded to each other, such as a monatomic noble gas. Van der Waals radii are typically much larger than covalent radii. Because of the lack of covalent compounds of neon, van der Waals radii are often used for this element as in Table 3.1. The estimated covalent radius of neon is, in fact, *slightly* larger than that of fluorine because of electron repulsions in the fully filled orbitals of the neon atom.

ii. A plot of first ionization enthalpies as we move across the second period of the periodic table is shown in Figure 3.2.

The overall trend in ionization enthalpy is that it **increases** as we move across a period. This again is a result of the increase of Z_{eff}.

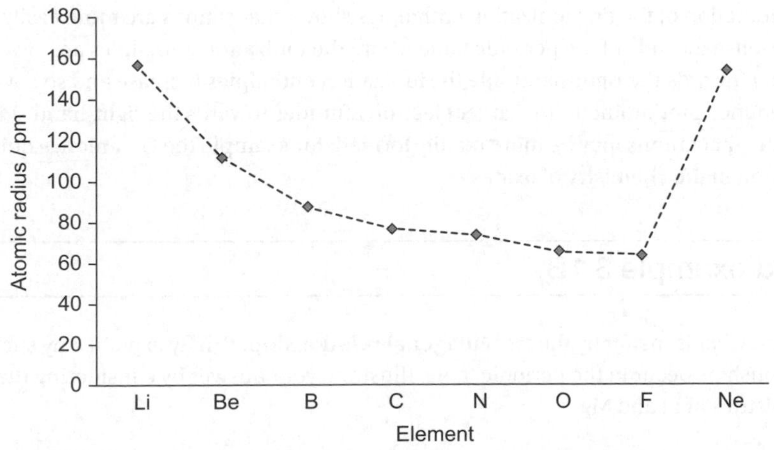

Figure 3.1 Atomic radii of the second row of the elements in the periodic table.

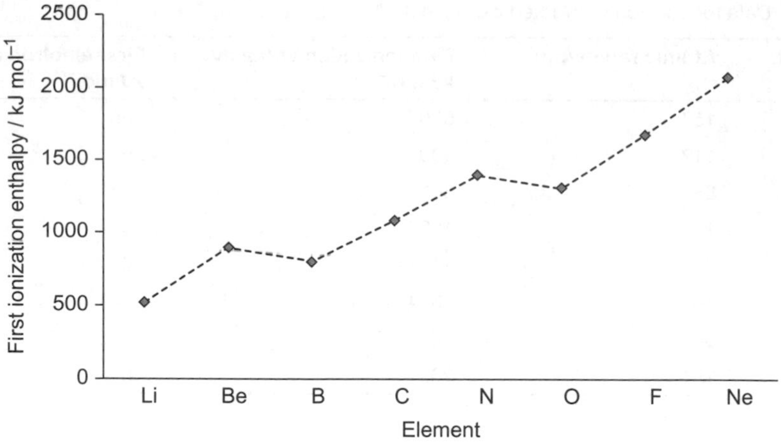

Figure 3.2 First ionization enthalpy of the second row of the elements in the periodic table.

This trend across a period may be contrasted with the situation seen as a group is descended. For example, on descending group 1 the ionization enthalpies **decrease** as follows: Li, 520; Na, 496; K, 419; Rb, 403; Cs, 376 kJ mol^{-1}. This trend may be explained because although shielding of the nuclear charge is always imperfect and thus Z_{eff} *increases* down the group, the outermost electron is, on average, further from the nucleus. These two effects act in opposition to each other. Interestingly, the first ionization enthalpy of francium (Fr) is *higher* than that of Cs. From Li to Cs the distance of the outermost electron from the nucleus is the dominant factor and so the first ionization enthalpy *decreases* as the group is descended. For Fr it is the increased value of Z_{eff} that dominates.

In comparing the ionization energies of N and O it is also important to consider *exchange energies*. An atom or ion is more stable if there are more *equivalent* ways to exchange the electrons. There are 3 equivalent ways to exchange the electrons in an N atom ($2s^2 2p^3$) as 3 electrons half-fill the three $2p$ orbitals. However there are only 2 equivalent ways to exchange the electrons in N$^+$ ($2s^2 2p^2$). By contrast there are 3 equivalent ways to exchange the electrons in *both* O ($2s^2 2p^4$) and O$^+$ ($2s^2 2p^3$) making O easier to ionize than N.

You should remember the rules for the filling of atomic orbitals with electrons. The important rule here is **Hund's rule of maximum multiplicity** which states that when filling **degenerate** orbitals—those with the same energy—the orbitals are first half filled by electrons with parallel spin, and when all have been half filled then they become fully filled as more electrons are added.

In determining the stability or otherwise of an ionic compound, it is important to consider a number of energy terms and not just a single term such as first ionization enthalpy or first electron affinity in isolation. This process requires the use of a **Born–Haber Cycle** which is discussed in more detail in Chapter 4.

There are two anomalies in this graph where a smooth straight line is not followed. First there is a decrease in the first ionization enthalpy between Be and B. This may be explained in terms of the **electronic configurations** of the two atoms. Be has the configuration $1s^2 2s^2$, while B has the configuration $1s^2 2s^2 2p^1$. Thus the electron lost from B is a $2p$ electron which is at higher energy than the $2s$ electrons and so more easily ionized. Moreover, the $2p$ electron is shielded by the filled $2s$ orbital. Secondly there is a decrease in first ionization energy between N and O. N has the electronic configuration $1s^2 2s^2 2p^3$ while O has the configuration $1s^2 2s^2 2p^4$. In N, all of the $2p$ electrons are unpaired, each one occupying one of the three $2p$ orbitals.

It requires less energy to remove an electron from an orbital where the electrons are paired because there is repulsion—known as the pairing energy—between the two electrons paired in a single orbital.

(b) The first electron affinities are negative (exothermic) for Li, B, C, O, and F but endothermic for Be, N, and Ne. This observation may be explained if it is recognized that in Be the extra electron is being added to a $2p$ orbital (which is shielded from the nuclear charge by the filled $2s^2$ orbital) and in Ne a new electron shell is being filled as the extra electron is being added to the $3s$ orbital. In N the extra electron must pair up in one of the $2p$ orbitals and, as already mentioned, this leads to repulsion because of the pairing energy. For Li, Be, C, and O the general trend is that the values become more exothermic as one moves from left to right across the period and this reflects the increasing effective nuclear charge (Z_{eff}).

(c) Consideration of the first ionization enthalpies shows that cations are more easily formed at the left-hand side of the periodic table where the ionization enthalpies are lower. On moving towards the right-hand side the ionization enthalpies increase and so covalent compounds predominate. The larger electron affinities towards the right-hand side indicate that anions may be more readily formed; for example the O^{2-} anion is quite common in the chemistry of oxides.

Worked example 3.1B

(a) Explain what is meant by the term **diagonal relationship**. Briefly explain why such relationships occur in the periodic table. Illustrate your answer by considering the chemistries of Li and Mg.

(b) Both potassium and copper have a single electron in a 4*s* orbital. Although potassium chemistry is dominated by K^+ cations and the +2 oxidation state is unknown, Cu(I) compounds are generally easily oxidized to Cu(II) and Cu(III) is known. Explain these observations in terms of the electronic configurations and ionization enthalpies of the two elements.

Solution

(a) A diagonal relationship occurs between an element and the element occupying the position immediately diagonally below and to the right. A very good example is the diagonal relationship between Li and Mg. As discussed in Worked example 3.1A, on moving from left to right across a period, atomic and ionic radii decrease while ionization enthalpies increase. On moving from the top to the bottom of a group, the atomic radius and ionic radius increases while the ionization enthalpy decreases. Thus on moving diagonally from Li to Mg the vertical and horizontal trends tend to cancel out and the two elements have similar properties.

Similarities in the chemistries of Li and Mg are:

i. Both metals (unlike the other group 1 metals) react with N_2 to form a stable nitride.

ii. Both metals burn in air to form oxides (unlike the other group 1 metals which form **peroxides** where the O—O bond remains intact).

iii. For both metals, the carbonates and nitrates decompose on heating to form oxides on heating.

iv. Both metals form a range of organometallic compounds which show a strong degree of covalency in the M—C (M=Li or Mg) bonds.

These properties may be considered in terms of the **polarizing** power of the cations—a property that depends to a large extent on the ionic radius. Ionic radii are determined from crystallographic experiments, and are based on the distances between neighbouring nuclei. As a result, ionic radii depend on the coordination of the ion (see Chapter 4 for more detail on the packing of ions in crystal structures). Mg^{2+} and six-coordinate Li^+ have very similar ionic radii of 72 and 76 pm respectively. The small size of both, and hence their high polarizing power, tends to stabilize nitrides and oxides and to give a degree of covalency (electron pair sharing) in the M—C bonds.

(b) Potassium has the electronic configuration [Ar]$4s^1$ while copper has the configuration [Ar]$3d^{10}4s^1$. The first second and third ionization enthalpies of potassium are: 418.8, 3052, and 4420 kJ mol^{-1} respectively. The corresponding enthalpies for copper are: 745.5, 1958, and 5536 kJ mol^{-1}. It takes more energy to remove the 4*s* electron of copper than it does to remove the corresponding electron of potassium. This reflects the poor shielding power of the 3*d* electrons. By contrast it is easier to remove a second or third electron from copper than it is from potassium. In copper these electrons are removed from the inner 3*d* orbitals; in potassium the electrons must be removed from the 3*p* orbitals, which is much less favourable.

Lithium and magnesium provide an excellent example of a diagonal relationship but other examples are known. Those between Be and Al and between Na and Ca are strong and weaker diagonal relationships exist between B and Si, C and P, and N and S.

Another factor that favours covalency is a high ionization enthalpy—removing electrons from metals such as lithium has a high energetic cost. This high cost also favours electron sharing.

These two elements, potassium and copper show just how important the filling of atomic orbitals is in influencing the chemical properties of an element. For example, copper is unreactive towards water and is often used to make water pipes. Potassium, by contrast, is very highly reactive towards water and yet both, as mentioned above, contain a single 4*s* electron.

Question 3.1

Look up the first ionization enthalpies for the elements of the *s* and *p* blocks of the third period of the periodic table, namely Na, Mg, Al, Si, P, S, Cl, and Ar. Plot these values and comment on i. the overall trends seen and ii. any specific anomalies that are seen in your graph.

Question 3.2

(a) For each of the following pairs of elements decide which has the higher first ionization enthalpy (you should be able to do this without looking the values up if you have worked through the example questions). Briefly explain your answer.

i. Rubidium and strontium.

ii. Barium and thallium.

iii. Strontium and barium.

(b) Account for the following trends in first ionization enthalpy:

i. The increase from bromine to krypton.

ii. The decrease from antimony to bismuth.

iii. The decrease from arsenic to selenium.

3.2 Group 1

The chemistry of the group 1 elements, which are also known as the **alkali metals**, is dominated by ions in the +1 oxidation state. These elements are highly reactive due to their relatively low enthalpies of atomization and ionization. Clear trends in the reactivity of these elements are usually observed, with notable exceptions for lithium.

The lithium cation (Li^+) is small hence has a high charge-to-size ratio, which makes it a **polarizing** cation. Furthermore, it has a high ionization enthalpy relative to the rest of group 1. These factors tend to favour a degree of covalent character in lithium compounds which is not seen with the other group 1 elements.

Worked example 3.2A

(a) Sketch a graph to show the trends in:

i. atomic radius and

ii. first ionization energy

for the group 1 alkali metals, as the group is descended. Explain the anomalies in these properties that occur with rubidium.

Solution

Depending on the information available to you, the graphs you produce can either be based on the actual numerical values for the quantities (atomic radius and first ionization energy) or a sketch of the rough trends observed, derived from your own knowledge.

Whilst the rest of this solution assumes that the actual quantities are available to you, the reasoning applied to the answer is applicable whether you have those details or not.

(a) First, if it has not been provided to you in the question directly, it is a good idea to look up the quantities required. Once the data have been obtained, construct a table of values, and plot the data.

Element	Atomic radius/pm	First ionization energy/kJ mol^{-1}
Li	145	520
Na	180	496
K	220	419
Rb	235	408
Cs	260	376

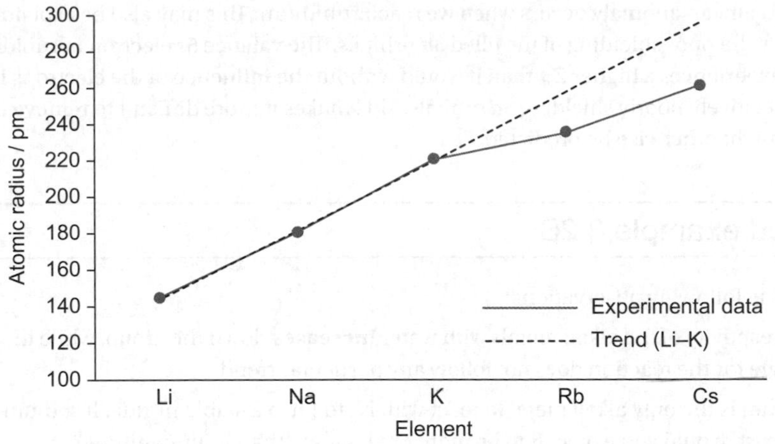

Figure 3.3 The atomic radius of the group 1 elements increases down the group.

i. The trends in atomic radius are illustrated in Figure 3.3.
As can be seen, the general trend is an increase in atomic radius as we descend the group. This can be explained by the increasing number of electrons contained in the neutral atoms, as we move down the group from lithium to caesium. An anomalous 'dip', relative to the dashed trend-line is observed for rubidium. This change may be explained by the filling of the 3*d* electrons, which are relatively poorly shielding. This poor shielding means that the outermost 5*s* electron in the rubidium atoms feels a greater effective nuclear charge (Z_{eff}) than would be expected were it not for the poor shielding properties of the 3*d* electrons. This increased attraction results in a smaller atomic radius than would be predicted (dashed line in the plot above) without knowledge of the poor shielding by the electrons in the 3*d* orbitals.

ii. The trends in first ionization enthalpy are illustrated in Figure 3.4.
The general trend in this case is straightforward to explain. The increasing number of electrons must occupy progressively higher energy orbitals as we descend the group. Electrons in higher energy orbitals spend more time further from the nucleus, and hence experience a lower degree of attraction. The outermost (valence) electron is shielded from the nuclear charge by the innermost electrons, although as mentioned previously such shielding is imperfect so Z_{eff} *increases* down the group. Overall these effects lead to a lower degree of attraction for the outermost electron down the group which results in its removal becoming progressively easier.

As discussed previously, the effective nuclear charge Z_{eff} *increases* as the group is descended because shielding of the nuclear charge is imperfect. The 3*d* electrons shield less well than *s* or *p* electrons. As such Z_{eff} for Rb and Cs is rather larger than might be predicted. This leads to smaller than predicted atomic radii (Figure 3.3) and higher than predicted ionization enthalpies (Figure 3.4).

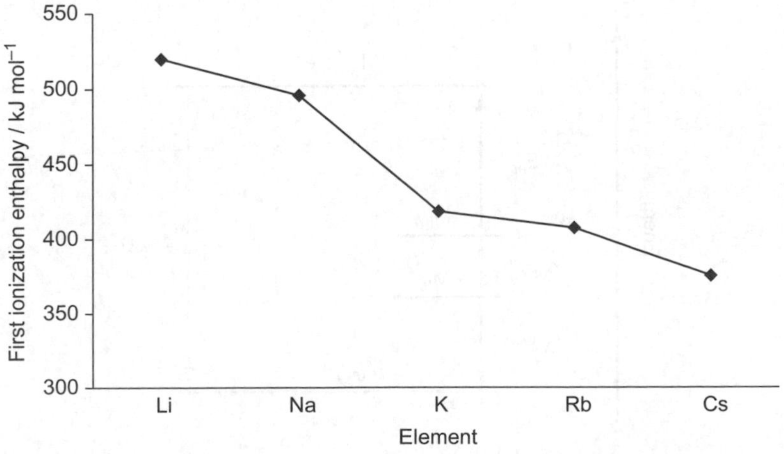

Figure 3.4 The first ionization enthalpy of the group 1 elements decreases down the group.

Again, an anomaly occurs when we reach rubidium. This may also be explained by the poor shielding of the filled 3*d* orbitals. The valence 5*s* electron in rubidium experiences a higher Z_{eff} than it would without the influence of the electrons in the relatively poorly shielding 3*d* orbitals. This makes it more difficult to remove than might otherwise be predicted.

Worked example 3.2B

Explain the following observations:

(a) The reactivity of the alkali metals with water **increases** down the group, while the enthalpy change for the reaction does not follow any particular trend.

(b) Lithium is the only alkali metal to react with N_2 to form a stable nitride. If sodium nitride did exist, would you expect it to be more or less ionic than lithium nitride?

Solution

(a) The alkali metals react with water according to the following general equation where M is an alkali metal:

$$M(s)+H_2O(l) \rightarrow M^+(aq)+OH^-(aq)+\frac{1}{2}H_2(aq)$$

A simple application of Hess's law allows us to break down the enthalpy change associated with the formation of M^+ (aq) ions into a series of theoretical steps. The enthalpy changes for these steps are generally known values that can be looked up, or are otherwise obtainable from experiment. These steps are summarized in the Hess's cycle shown in Figure 3.5.

The enthalpy changes associated with atomization and ionization are positive (endothermic) terms, i.e. energy is required for these steps. This energy requirement contributes towards the **activation energy** for the reaction. The activation energy for a reaction is a thermodynamic barrier that must be overcome to enable a reaction to proceed. A good explanation can be found in section 9.7 of Burrows et al. (2013).

As we descend the group, the increased size of the atoms results in weaker metallic bonding. The atoms in a metal are held together by the attraction between the positive nuclei of the atoms, and a delocalized 'sea' of electrons, which are distributed over the whole of the metal sample. As the atoms become progressively larger, the distance between the valence electrons and the nuclei increases. This results in less attraction between the nuclei and the electrons,

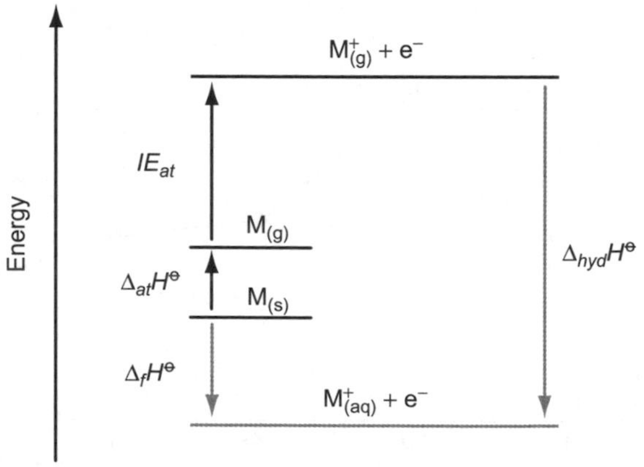

Figure 3.5 Hess's cycle showing how the enthalpy of formation of the M^+ (aq) cation is related to known experimental values.

making it easier to break apart the metal. This results in a decrease in the enthalpy of atomization down the group. Similarly, the first ionization enthalpy decreases down the group. This is due to the outermost *s* electron occupying an orbital with an increasing value of *n*, and hence higher energy as the group is descended. This electron is on average further from the nucleus than the equivalent valence electron in the element preceding it in the group, making it easier to remove. The combination of the decreasing enthalpies of atomization and ionization result in the increased reactivity observed as we move down the group.

However, the reactivity does not tell us anything about the overall enthalpy change for the reaction, as we have not yet discussed the enthalpy of hydration. As seen in the diagram above, the energy put into the system is recovered with the enthalpy of hydration, which depends on the size of the cation. Hydration is an exothermic process, as new bonds are formed between the cation and the water molecules. More polarizing cations form stronger bonds with water, and hence smaller cations result in increasingly negative hydration enthalpies (i.e. more energy is given out). This means that while the barrier to enable a reaction to occur (activation energy) decreases down the group, the amount of energy given out by the reaction also decreases down the group. It is these two factors combined that result in the seemingly trendless values observed for the enthalpy changes when alkali metals react with water.

This question illustrates why it is often important in chemistry to consider a number of factors that may affect the outcome of a reaction. In this case, it is clear that while the reactivity trend for a group may be predicted, that does not necessarily allow us to predict the trend for the enthalpy change for the reaction as we descend the group.

(b) Lithium forms a stable nitride while the rest of the group 1 elements do not. This may be rationalized by the small size of the Li^+ ion. The small size of the Li^+ cation, and the N^{3-} anion means that the product has a very high lattice enthalpy. This lattice energy is sufficiently large that even the strong covalent N≡N bond can be broken and still result in an overall exothermic reaction. The lattice enthalpies for the other alkali metal nitrides are not sufficiently large to enable an exothermic reaction, and hence no other stable alkali metal nitrides are known.

The high charge-to-size ratio of the lithium cation means it is a **polarizing** cation. The polarizing power of the Li^+ cation distorts the electron density around the N^{3-} anion, introducing a degree of covalent character to the bond. The Na^+ cation has the same charge as Li^+ but is larger, and therefore has a lower charge-to-size ratio. This means it is less able to distort the electron density of the anion, resulting in a lower degree of covalency in the product.

Charge-to-size ratio is sometimes referred to as *charge density* but *charge-to-size* ratio is used throughout in this book.

Figure 3.6 shows how the lithium cation distorts the electron cloud around the nitride anion to a greater degree than the equivalent sodium cation. This means that we would expect sodium nitride to be more ionic than lithium nitride.

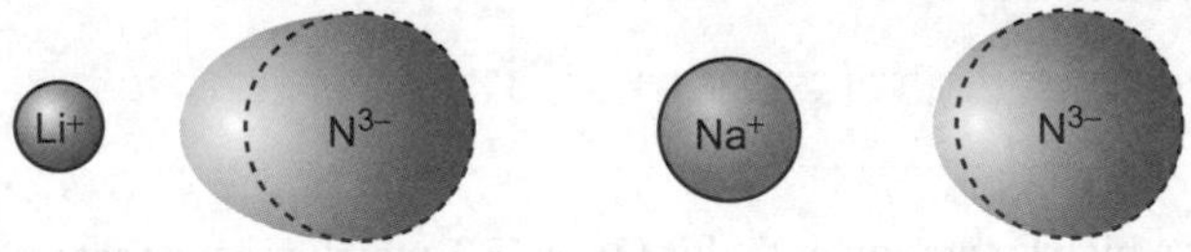

Figure 3.6 Highly polarizing cations such as Li^+ distort the electron clouds of neighbouring anions more than less polarizing cations, such as Na^+. The greater electron density between the two ions results in a degree of covalent character in the bonding.

Question 3.3

Although the ionization potentials for oxidation of the group 1 metals to their cations decrease as the group is descended, the standard electrode potentials (E°) for the process:

$$M^+(aq) + e^- \rightarrow M(s)$$

remain relatively constant down the group, as shown in the following table. Explain this observation.

Metal	$E^{\ominus}$/V
Li	−3.04
Na	−2.71
K	−2.93
Rb	−2.92
Cs	−2.92

Question 3.4

Explain why most of the group 1 metals must be obtained by the electrolysis of molten salts rather than from aqueous solutions.

3.3 Group 2

The chemistry of the group 2 elements, also known as the **alkaline earth metals**, is dominated by ions in the +2 oxidation state. These elements tend to be less reactive than their neighbours in group 1, with a number of important differences in their reactivity trends.

Like lithium in group 1, beryllium exhibits some anomalous behaviour with respect to the reactivity of the rest of the group. Again, this is explained by the high enthalpy of formation and highly polarizing nature of the Be^{2+} cation.

Worked example 3.3A

(a) Briefly explain why compounds of beryllium are mainly covalent whereas compounds of the other group 2 elements are mainly ionic.

(b) Sketch the structures adopted by the compounds $BeCl_2$ and BeH_2 in the solid state. Describe the difference between the gas-phase and solid-state structures, and describe the bonding around the beryllium atom in the solid state. Explain how, in each case, this bonding is covalent, and state clearly if either of these compounds is **electron deficient**, explaining your answer.

Solution

(a) Beryllium has a high ionization enthalpy ($IE_1 + IE_2$), which means that removing the valence electrons to form the Be^{2+} cation requires a great deal of energy. The Be^{2+} cation is small and highly charged and hence has a high charge-to-size ratio. Ions that have high charge-to-size ratios are said to be highly **polarizing**. This polarizing power favours a high degree of covalency in resultant compounds. In comparison, the rest of the alkaline earth metals have greater numbers of electrons, which consequently results in larger ionic radii and smaller ionization enthalpies, as shown by the data in Table 3.2. As a result, their propensity to form ionic compounds is greater than that of beryllium.

(b) $BeCl_2$ and BeH_2 form polymeric compounds in the solid state, as depicted in Figures 3.7 and 3.8 respectively. In the gas phase, both $BeCl_2$ and BeH_2 form linear molecules, with standard covalent bonds between the central beryllium and the bonded chlorine or hydrogen atoms. However, in the solid state, in both compounds, the beryllium is four-coordinate.

Table 3.2 Data used in the solution for Worked example 3.3A.

Element	IE_1 / kJ mol^{-1}	IE_2 / kJ mol^{-1}	$IE_1 + IE_2$ / kJ mol^{-1}
Be	899	1757	2656
Mg	738	1450	2188
Ca	590	1145	1735
Sr	549	1064	1613
Ba	503	965	1468

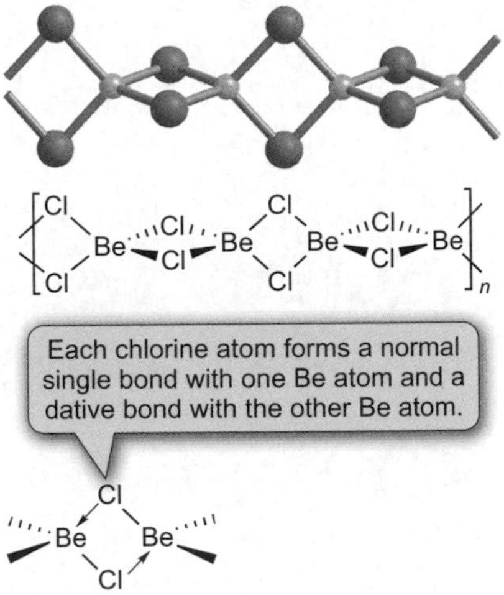

Each chlorine atom forms a normal single bond with one Be atom and a dative bond with the other Be atom.

Figure 3.7 The structure of $BeCl_2$ contains bridging chlorine atoms.

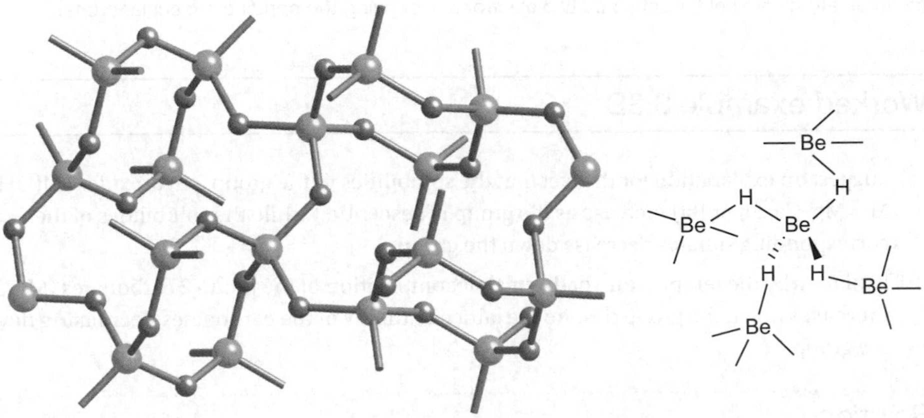

Figure 3.8 Solid BeH_2 forms a three-dimensional polymeric structure with bridging hydrogen atoms.

In $BeCl_2$, the bridging chlorine atoms donate three electrons each: one donates an electron pair with a beryllium valence electron to form a standard covalent bond, while a lone pair of electrons on the chlorine atom is donated to a neighbouring beryllium atom to form a dative covalent (coordinate) bond.

Hydrogen does not have any lone pairs available to form dative covalent bonds and so it is not possible to draw a traditional Lewis dot-and-cross diagram for the bonding observed. Instead, beryllium hydride is an electron-deficient compound that features three-centre, two-electron (3c2e) bonding.

By making use of sp^3 hybrid orbitals on the beryllium atoms, Figure 3.9 shows how these can overlap with the hydrogen 1*s* orbital to give bonding, non-bonding, and antibonding orbitals.

Each Be atom has two electrons which are shared between four sp^3 hybrid orbitals. Given that half of the sp^3 orbitals on each beryllium atom are empty, each 3c2e bond therefore contains one electron from the bridging hydrogen atom, and one electron from one of the two Be atoms, as shown in Figure 3.10.

The pair of electrons occupies a bonding orbital, giving rise to a 3c2e bond, which is also known as a 'banana bond'. Molecular orbitals are covered in more detail in Chapter 2 of this workbook.

In both cases, electrons are shared between atoms rather than fully transferred meaning both compounds exhibit covalent bonding.

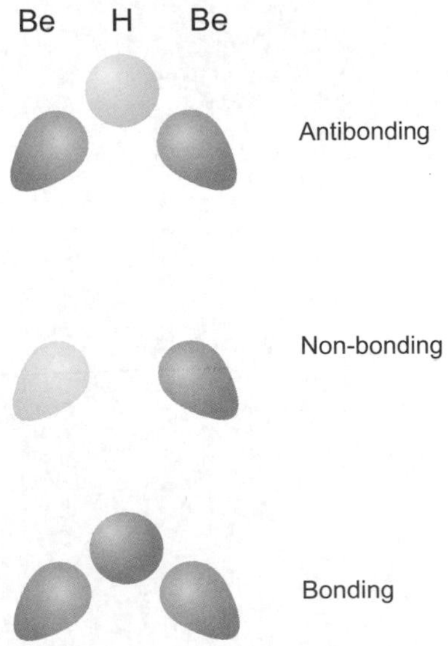

Figure 3.9 Two sp^3 hybrid orbitals on two beryllium atoms can combine with the 1*s* orbital on a bridging hydrogen atom to form bonding, non-bonding, and antibonding molecular orbitals.

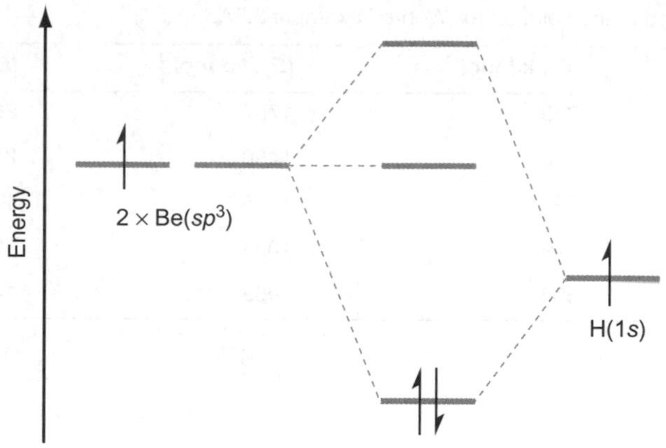

Figure 3.10 Molecular orbital energy level diagram for a Be–H–Be unit in solid BeH_2, showing how the three atoms are held together by two electrons occupying the bonding molecular orbital.

Worked example 3.3B

(a) Suggest an explanation for the fact that the solubilities of the group 2 hydroxides ($M(OH)_2$, M = Mg, Ca, Sr, or Ba) increase as the group is descended while the solubilities of the corresponding sulfates decrease down the group.

(b) Explain why the temperature of thermal decomposition of the group 2 carbonates (MCO_3) **increases** down the group despite the lattice enthalpy of the carbonates **decreasing** down the group.

Solution

(a) At a simplistic level, the solubility of a given salt in water can be understood as a balance between the lattice enthalpy of dissociation of the salt (i.e. the energy required to break apart the solid lattice), and the sum of the enthalpies of hydration for the constituent ions (i.e. the enthalpy change when the ions enter solution); this situation is depicted in Figure 3.11.

A salt with a high lattice enthalpy favours remaining in the solid state due to the high energetic cost of breaking the ionic lattice. A salt with constituent ions that have high enthalpies of hydration favours dissolution, in that the enthalpy required to break the ionic lattice is compensated for by the hydration enthalpy. In a real salt, these factors compete with one another, and the resultant trends are observed.

In the example given in the question, the solubility of the group 2 metal hydroxides increases down the group. This can be explained in terms of the decreasing lattice enthalpy down the group. The increased size of the ions down the group results in lower lattice enthalpies, meaning the enthalpy required to break the lattice decreases, and dissolution is favoured.

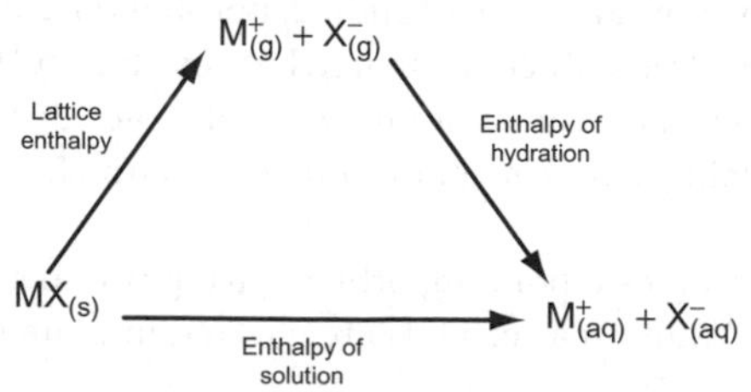

Figure 3.11 A good starting point for understanding the solubility of a given salt is to look at the competing factors of lattice enthalpy and hydration enthalpy.

For the sulfates, the opposite trend is observed. The decreasing solubility down the group may be explained in terms of the decreasing hydration enthalpy of the metal ions down the group.

A comparison of the lattice enthalpies of the sulfates and hydroxides is not simple. The sulfate ion is larger than the hydroxide ion and this tends to make the lattice enthalpy *smaller*. However, the sulfate ion is doubly charged and this tends to make the lattice enthalpy *larger*.

Whilst this explanation is relatively straightforward, the competing factors of hydration enthalpy and lattice enthalpy mean that it is often difficult to predict solubility trends.

One final point to consider here is that in many examples, the enthalpy change for dissolving a salt is positive (i.e. it is an endothermic process). Why then does the reaction occur spontaneously? To understand that, we must consider not just the enthalpy change but also the entropy change associated with the process; breaking the highly ordered crystalline lattice results in a strong positive entropy change. Therefore, we must study the process in terms of the Gibbs energy change:

$$\Delta G = \Delta H - T\Delta S$$

If the term $T\Delta S$ is sufficiently large, processes with endothermic enthalpy changes can still result in a negative value for ΔG and therefore still occur spontaneously.

As noted it is difficult to predict the trend in lattice enthalpies of sulfates and hydroxides because of the competing effects of size and charge. Interestingly, experimental lattice enthalpies show that $Ca(OH)_2$ has a larger lattice enthalpy than $CaSO_4$, $Sr(OH)_2$ has a lattice enthalpy very similar to that of $SrSO_4$, while $Ba(OH)_2$ has a *smaller* lattice enthalpy than $BaSO_4$. Values for calculated lattice enthalpies (kJ mol^{-1}) are: $Ca(OH)_2$, 2645; $CaSO_4$, 2480; $Sr(OH)_2$, 2483; $SrSO_4$, 2484; $Ba(OH)_2$, 2339; $BaSO_4$, 2374.

(b) The lattice enthalpy of the group 2 metal carbonates decreases down the group, due to the increasing size of the metal cation. It would be expected from this trend that the thermal decomposition temperature of the carbonates should therefore also decrease down the group, due to the ionic lattices becoming increasingly easy to break apart. However, the opposite trend is seen.

The general equation for this reaction is as follows:

$$MCO_3(s) \xrightarrow{\Delta} MO(s) + CO_2(g)$$

The trend observed can be explained by considering the lattice enthalpy of the metal oxide product formed. The lattice enthalpy of the MO product decreases down the group. The formation of the MO lattice in part helps to offset the energy needed to break down the MCO_3 lattice. As such, the decreasing 'pay back' obtained from the formation of MO down the group means that the reaction becomes less favourable, and hence requires a greater energy input.

There is a large favourable entropy change in this reaction which is caused by the production of CO_2 gas as a product.

Question 3.5

Give chemical equations for the thermal decomposition of the

(a) sulfates and

(b) hydroxides

of the group 2 metals, and predict the order of thermal stability of these salts on descending group 2. You may wish to consider your answer to Worked example 3.2B.

Question 3.6

Explain why the group 2 metals form stable nitrides but do not form stable peroxides (containing the O_2^{2-}) ion.

3.4 **Group 13**

Group 13 is dominated by **covalent** chemistry with the group oxidation state predominating. The principal anomalies in this group occurs at the bottom of the group. Thallium in many ways is much more akin to the group 1 metals potassium or rubidium than it is to the other members of group 13. It is often found as a +1 cation in its compounds and it forms a wide range of salts which are isostructural with those of potassium, rubidium, and caesium (such as the thallium halides, which, apart from the fluoride, are isostructural with caesium chloride). The reason for this behaviour rests on the **inert pair effect** where the pair of 6*s* electrons on thallium **penetrate** close to the nucleus, which for this heavy metal has a high charge. The high charge on the nucleus and the penetration of the 6*s* electrons leads to a high ionization enthalpy for these electrons. As such, this pair of electrons behaves as an **inert pair** that does not readily become involved in chemistry. At the same time, the $6p^1$ electron penetrates less to the nucleus and is relatively readily ionized. As such, a range of thallium salts with the Tl^+ ion are well known. A similar effect is seen in group 14 where lead and tin show chemistry of the M^{2+} ion (M = Sn or Pb) in many of their compounds.

Worked example 3.4A

Comment on the following observation. The most stable halides of boron and aluminium are the covalent trihalides whereas for thallium, the ionic monohalides (TlX, where X = halogen) are the most stable. Similarly, the most stable oxides of boron and aluminium have the formula M_2O_3 (M = B or Al) whereas the most stable oxide of thallium has the formula Tl_2O.

Solution

This is an example of the inert pair effect, which is described in the introduction to this section. As hinted at in the question, thallium chemistry is dominated by the +1 oxidation state and in many respects it resembles a group 1 alkali metal rather than a group 13 element. In this case we see thallium forming ionic halides and oxide. By contrast boron forms a monomeric covalent trihalide with a trigonal planar structure, while aluminium chloride, $AlCl_3$, exists at room temperature as a dimer Al_2Cl_6 where two $AlCl_3$ units are held together by dative covalent (coordinate) bonding from bridging Cl atoms; these structures are shown in Figure 3.12.

B_2O_3 typically forms an amorphous (i.e. glassy) solid that can be crystallized after prolonged exposure to heat, while Al_2O_3 crystallizes into a number of well characterized forms, one of which forms the basis of gemstones such as ruby and sapphire.

The colour of the gemstones ruby and sapphire results from impurities in the crystal structure of pure Al_2O_3, which is itself colourless. The red colour of ruby stems from chromium impurities, while the various colours of sapphire result from trace quantities of copper (orange), iron (blue), magnesium (green), and titanium (yellow).

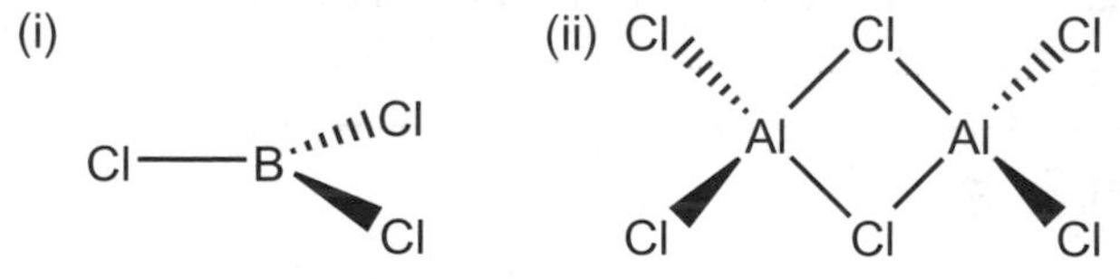

Figure 3.12 The structures of (i) BCl_3 and (ii) Al_2Cl_6.

Worked example 3.4B

For each of the reductions given in the table below calculate a value for $\Delta G^{\ominus}$ using the equation:

$$\Delta G^{\ominus} = -nFE^{\ominus}$$

(where F = Faraday constant, 9.6485×10^4 C mol^{-1} and n is the number of electrons involved in the redox reaction).

Hence determine standard free energies of reaction ($\Delta G^{\ominus}$) for the reaction of Tl metal and dilute acid to produce i. Tl^+ (aq) and H_2 (g) and ii. Tl^{3+} (aq) and H_2 (g). Hence determine which of the reactions i. or ii. is thermodynamically more favourable and comment on your answer.

Half-equation	$E^{\ominus}$/V
$Tl^+(aq) + e^- \rightarrow Tl(s)$	−0.33
$Tl^{3+}(aq) + 3e^- \rightarrow Tl(s)$	0.72
$2H^+(aq) + 2e^- \rightarrow H_2(g)$	0.00

Solution

This part of the question explores the ground that the previous part covered but quantitatively. What are the standard Gibbs energy changes associated with the oxidation of thallium to the +1 or +3 oxidation states?

Standard Gibbs energies ($\Delta G^{\ominus}$) may be calculated for the standard electrode potential ($E^{\ominus}$) by:

$$\Delta G^{\ominus} = -nFE^{\ominus}$$

(where F is the Faraday constant = 9.6485×10^4 C mol^{-1}).

We can therefore calculate $\Delta G^{\ominus}$ values for the overall reactions:

Reaction	$\Delta G^{\ominus}$/kJ mol^{-1}
$2Tl(s) + 2H^+(aq) \rightarrow 2Tl^+(aq) + H_2(g)$	−63.6
$2Tl(s) + 6H^+(aq) \rightarrow Tl^{3+}(aq) + 3H_2(g)$	+416

Thus it may be concluded that thallium metal will react with acid to give hydrogen gas (as expected for a metal with a negative standard electrode potential, and hence higher in the reactivity series than hydrogen). However, it may be seen that the stable product is Tl^+(aq) and not Tl^{3+}(aq). The formation of Tl^+(aq) has a negative (favourable) standard Gibbs energy change while the formation of Tl^{3+}(aq) has a positive (unfavourable) standard Gibbs energy change. These calculations are designed to back up the qualitative description that was considered in Part (a).

Thus for the three processes listed in the table in the question, the $\Delta G^{\ominus}$ values are:

+31.8 kJ mol^{-1};
−208 kJ mol^{-1};
0 kJ mol^{-1}.

Question 3.7

The ordering of the relative stabilities of adducts L-BH_3 for some common Lewis-base donor ligands is, according to L: $Et_2O < Me_2S < Me_3N < Me_3P < H^-$.

State whether or not you would expect the Et_2O ligand to be displaced when i. Me_3N and ii. $Et_2PCH_2CH_2PEt_2$ is added to an Et_2O solution of $Et_2O.BH_3$. What are the possible stoichiometries of the product(s) formed?

You should remember that a Lewis base is a molecule or ion that can donate a pair of electrons to an acceptor species. The species which accepts the pair of electrons is known as a Lewis acid.

Question 3.8

Describe the bonding in i. Ga_2H_6 and ii. Ga_2Cl_6.

► **Hint** Both of these materials have structures similar to that of Al_2Cl_6.

3.5 Group 14

Remember, the **inert pair effect** arises from the high degree of penetration of the 6*s* electrons, and the high nuclear charge seen with metals at the bottom of the group. These electrons are held closely by the nucleus and hence are relatively difficult to remove.

Group 14 is dominated by **covalent** chemistry with the group oxidation state predominating. The principal anomalies in this group occurs at the bottom of the group. As is seen in group 13 with thallium, the **inert pair effect** is also seen in group 14, where lead and tin show chemistry of the M^{2+} (M = Sn or Pb) in many of their compounds.

Worked example 3.5A

Comment on the following observations.

(a) Under ambient conditions, carbon dioxide exists as a gas whose molecules feature π bonding whereas silicon dioxide is a solid with σ bonds between Si and O, and has a very high melting point.

(b) Carbon is limited to a valency of four. However, the heavier elements in this group readily form a series of complexes $[MF_5]^-$ where M = Si, Ge, Sn, or Pb.

Solution

This question explores some of the typical trends that are seen on descending group 14 and contrasts the well-known chemistry of carbon with the perhaps rather less familiar chemistry of the heavier elements.

Part (a) considers p_π—p_π bonding (see Chapter 2). This is a well-known feature of the chemistry of carbon and there are many examples of carbon compounds which show this type of bonding. CO_2 is a good example. Thus CO_2 exists at room temperature as discrete covalent molecules and is gaseous. As the group is descended, the π orbitals become more diffuse, the overlap weaker, and hence the pπ bonds become weaker. As such it becomes energetically more favourable to form four σ bonds than to form two π bonds. Silicon–oxygen σ bonds are particularly strong and are a very common feature of silicon chemistry. SiO_2 forms a giant covalent structure and as such has a very high melting point.

This situation is analogous to three-centre, two-electron bonding (see Chapter 2). However, in three-centre, two-electron bonds, only the lowest energy bonding molecular orbital is filled. In three-centre, four-electron bonds, the bonding and non-bonding orbitals are filled. In both cases, two bonding electrons hold together three atoms.

Part (b) looks at the coordination numbers seen on descending the group. Carbon is limited to a valency of four and an 'octet' of electrons in its compounds. As the group is descended, larger coordination numbers are seen and expansion of the 'octet' of electrons is formally observed. An important point here is simply size—a carbon atom is smaller than atoms of the other elements in the group (atomic radii: C = 70 pm, Si = 110 pm, Ge = 125 pm, Sn = 145 pm, Pb = 180 pm). In the past, the role of *d* orbitals in expanding the octet of electrons has been considered. However, theoretical calculations suggest that the role of *d* orbitals is not so important and that instead, three-centre, four-electron (3c4e) bonds may form, where one pair of electrons is in a non-bonding orbital. This allows the octet to expand—but only if the central atom is large enough.

Worked example 3.5B

Comment on the following observations:

(a) Carbon tetrachloride (CCl_4) is a stable molecule whereas the dichloride (CCl_2) exists only as an unstable intermediate. By contrast, lead tetrachloride ($PbCl_4$) readily decomposes on heating at 50 °C to give a stable dichloride ($PbCl_2$).

$$PbCl_4 \rightarrow PbCl_2 + Cl_2$$

(b) Carbon readily **catenates**, i.e. forms compounds with long chains of carbon atoms, in some cases many thousands of carbon atoms long. By contrast such chain molecules become much less stable down group 14—silicon forms chains with a maximum of 12 atoms and the maximum number of tin atoms known to bond together is three.

Solution

(a) The **inert pair effect**, which has already been encountered in section 3.3, is a result of the fact that in lead, the $6s^2$ electrons are tightly held to the nucleus. This means that they do not easily engage in bonding which makes the +2 oxidation state relatively more stable. An additional argument is the relative strengths of C—Cl and Pb—Cl bonds. Pb—Cl bonds are significantly weaker (ΔH(C—Cl) = 339 kJ mol^{-1}; ΔH(Pb—Cl) = 243 kJ mol^{-1}) so as well as being easier to break apart, the lower bond enthalpy means that the formation of a lead tetrahalide is less favoured than the carbon tetrahalide.

(b) Catenation is a well-recognized feature of carbon chemistry and reflects the fact that C—C bonds are quite strong. The bond strength between atoms of group 14 elements falls off rapidly down the group as atomic radius increases and hence while silicon shows some propensity to catenate (though the size of the molecules is very limited when compared to carbon), for tin, catenation is almost unknown.

Bond	Bond dissociation enthalpy/kJ mol^{-1}
C—C	347
Si—Si	295
Sn—Sn	151

Question 3.9

Explain what is meant by the **inert pair effect** with reference to the chemistry of the *p*-block elements. Explain which electrons are considered to be **inert** when **divalent** compounds of group 14 elements are formed. Suggest an explanation for the fact that while divalent compounds of carbon are unstable, those of lead and tin are stable and that Pb^{4+} can act as an oxidant.

$$Pb^{4+} + 2e^- \rightarrow Pb^{2+} \qquad E^{\ominus} = +1.69 \text{ V}$$

Question 3.10

(a) Describe with the aid of diagrams the structures adopted by the tin chlorides $SnCl_4$ and $SnCl_2$ in their standard states.

(b) Write equations for the reactions of SiH_3Cl with i. NH_3, ii. H_2O vapour, and iii. Na metal. Comment on the structures of the reaction products.

3.6 Group 15

The chemistry of group 15 is dominated by covalent chemistry. However, while the ionization enthalpies of these elements, which lie to the right-hand side of the periodic table, are too high to readily allow the formation of cations, some anionic chemistry is observed at the head of the group where ionic nitrides (N^{3-}) are seen. The formation of anions in these cases is typically favoured by the small size of the resultant anion giving rise to a high lattice enthalpy for the product.

Worked example 3.6A

This question addresses the chemistry of phosphorus halides in a problem-solving manner. It is designed to use experimental data to determine the identities of a range of phosphorus compounds and to use this information to help understand the chemical reactivity of these compounds, the likely expected oxidation states and the structures that these compounds adopt.

Compound **A** is a phosphorus chloride that only contains phosphorus and chlorine. It contains 77.4% Cl by mass. It is a colourless liquid, with covalent bonding in the constituent molecules.

Compound **A** reacts with Cl_2 to give compound **B**, which is a second phosphorus chloride that again only contains phosphorus and chlorine. It is an ionic solid that contains 85.1% Cl by mass.

Compound **A** reacts with an excess of oxygen gas to form **C**. Compound **C** contains phosphorus, oxygen, and chlorine, and contains 69.4% Cl by mass. It is a colourless liquid with covalent bonding in the constituent molecules.

Compound **C** shows an infrared absorption at 1290 cm^{-1} whereas **A** and **B** show no infrared absorptions above 700 cm^{-1}.

(a) Give the empirical formula for **A**.

(b) Use VSEPR rules to predict the molecular structure of **A**.

(c) Give the empirical formula of **B**.

(d) Give a balanced equation for the formation of **B** from **A**.

(e) Give a possible structure for **B** showing the ionic nature of **B**.

(f) Give the empirical formula of **C**.

(g) Give a balanced equation to show the formation of **C** from **A**.

(h) Suggest which vibration in **C** gives rise to the infrared band at 1290 cm^{-1}.

Solution

(a) The question tells us that it is covalently bonded but this could be deduced by the fact that it is a liquid, i.e. it has a low melting point. The low melting point indicates that the compound is likely to be monomeric.

We can obtain the amount in moles of phosphorus and chlorine (per hundred grams of sample) by dividing the percentage by mass of chlorine and phosphorus in the sample by the molar mass of phosphorus and chlorine. This then allows us to determine the empirical formula by determining the ratio of these values.

Compound A	P	Cl
Mass %	22.6	77.4
Relative atomic mass	30.97	35.45
Moles/100 g	0.73	2.18
Empirical formula	1	3

Given that it is likely to be monomeric, we can deduce that the molecular formula of this material matches the empirical formula and therefore compound **A** is PCl_3, where phosphorus is in the +3 oxidation state.

This kind of question is solved by combining numerical calculations with chemical knowledge. In this case either prior knowledge or material obtained from outside reading would indicate that PCl_3 is a well-known phosphorus chloride.

(b) As the phosphorus atom is formally in the +3 oxidation state it has a lone pair so it has a pyramidal structure:

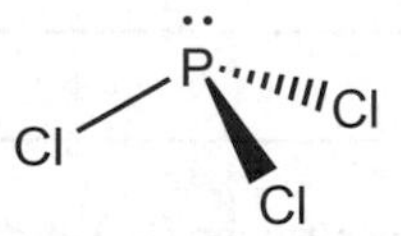

(c) Using a similar approach to part (a), we can construct a table to determine the empirical formula of **B.**

Compound B	**P**	**Cl**
Mass %	14.9	85.1
Relative atomic mass	30.97	35.45
Moles/100 g	0.48	2.40
Empirical formula	1	5

The data allow us to calculate that **B** has the formula PCl_5. This is relatively easy to predict in advance, as we expect phosphorus to exist in the +3 or +5 oxidation states.

(d) The chemical equation here is trivial but it is good and useful practice always to give a chemical equation for a chemical reaction, however simple this equation may seem. Clearly this is an oxidation reaction where phosphorus in the +3 oxidation state is oxidized by Cl_2.

State symbols should be included, as we are told that **A** is a liquid and **B** is a solid.

$$PCl_3(l) + Cl_2(g) \rightarrow PCl_5(s)$$

(e) Interestingly this is an **ionic** solid. Its solid state at room temperature can be explained by the fact that it contains ions. However we know that P^{5+} ions would be impossible to form—the enthalpy of formation for P^{5+} is about 17,000 kJ mol^{-1}. Thus it must contain **complex** molecular ions. In fact its structure is:

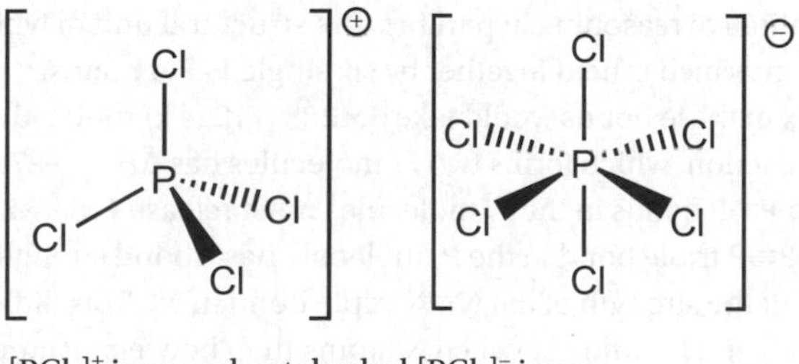

with tetrahedral $[PCl_4]^+$ ions and octahedral $[PCl_6]^-$ ions.

Without prior knowledge of this specific example, a sensible (though in this case, incorrect) answer would be:

$$[PCl_4]^+Cl^-$$

This structure is, in fact, adopted by the bromide PBr_5, i.e. it forms the ionic solid $[PBr_4]^+$ Br^-. This is presumably because six of the larger Br^- ions cannot easily fit around the central P atom to form $[PBr_6]^-$.

(f) In this case it is not possible to work out the formula directly from the data because there are two unknowns (the amounts of P and O). However, given chemical knowledge about the oxidation states of phosphorus then a very likely molecule can be predicted: $OPCl_3$. This can then be confirmed by calculating the percentage of Cl in the compound.

(g) This is another oxidation reaction in which phosphorus is oxidized from +3 to +5 oxidation states with O_2 acting as the oxidant in this case.

$$PCl_3(l) + \tfrac{1}{2}O_2(g) \rightarrow OPCl_3(l)$$

This can be shown by using the following equation, which calculates the **wavenumber** of a molecular vibration:

$$\bar{\nu} = \frac{1}{2\pi c}\sqrt{\frac{k}{\mu}}$$

where c is the speed of light, k is the force constant for the bond, and μ is the **reduced mass** of the two atoms in the bond, given by the equation:

$$\mu = \frac{m_1 m_2}{m_1 + m_2}$$

As can be seen from these equations, large values of k (force constant), or small values of μ (reduced mass) are likely to result in a correspondingly large wavenumber.

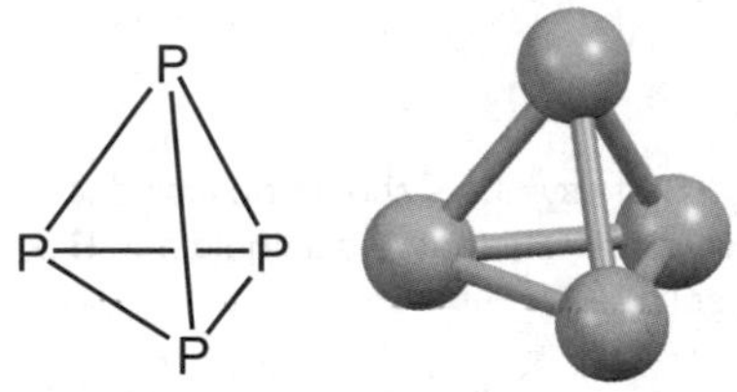

Figure 3.13 White phosphorus is composed of P_4 tetrahedra.

(h) The most likely candidate for this is the P═O bond as this is the only structural unit present in **C** which is not found in **A** or **B**. A useful rule of thumb in interpreting infrared spectra is that bands occurring above 1000 cm^{-1} either occur from bonds including a hydrogen atom (low reduced mass) or from double or triple bonds (high force constant).

Worked example 3.6B

(a) Contrast the structures adopted by nitrogen (N_2) and by the white allotrope of phosphorus (P_4).

(b) The decomposition reaction of gaseous tetrahedral P_4:

$$P_4(g) \rightarrow 2P_2(g)$$

has $\Delta H^{\ominus} = +274$ kJ mol^{-1}. Given that the bond enthalpy of a single P—P bond is 209 kJ mol^{-1} calculate the energy of the P≡P triple bond in P_2. Compare the value you obtain with the bond energy in N_2 (946 kJ mol^{-1}) and comment on any difference.

Solution

(a) N_2 shows strong π bonding and as such π-bonded molecules are stable. Phosphorus shows very complex allotropy; 12 allotropes (both amorphous and crystalline) have been reported. A common allotrope is white phosphorus which contains P_4 tetrahedra; its structure is depicted in Figure 3.13.

The P–P bond distances in white phosphorus of 221 pm are consistent with single σ bonds, as the covalent radius of the P atom is 110 pm. Table 3.3 shows the covalent bond enthalpies for single (σ) and triple ($\sigma + 2\pi$) NN and PP bonds. It may be seen that the single bond strength follows the order PP>NN whereas the triple bond strength follows the order NN>>PP. This observation reflects the decrease in the strength of π bonds as the group is descended as a result of the poorer overlap of the p_x and p_y orbitals.

(b) The second part of this question provides some quantitative thermodynamic data to back up the more qualitative reasoning in part (a). The structural unit of white phosphorus is the P_4 tetrahedron, which is held together by six single P–P σ bonds.

To break all six of these bonds would take $6 \times 209 = 1254$ kJ mol^{-1}. The enthalpy change for this reaction, which forms two P_2 molecules has $\Delta H^{\ominus} = +274$ kJ mol^{-1}, so the formation of two P≡P bonds in the P_2 molecules must release $1254 - 274 = 980$ kJ mol^{-1}. Therefore, each P≡P triple bond in the P_2 molecule has a bond strength of 490 kJ mol^{-1}. This is around half the strength of the N≡N triple bond in N_2. This difference reflects the significantly stronger π bonding between N atoms than between P atoms, which results from the more diffuse nature, and hence poorer overlap, of the $3p$ orbitals on phosphorus when compared to the $2p$ orbitals on nitrogen.

Table 3.3 Covalent bond enthalpies in kJ mol^{-1} for single and triple NN and PP bonds.

Bond	Bond enthalpy/kJ mol^{-1}
N—N	160
N≡N	946
P—P	209
P≡P	490

Question 3.11

(a) Sketch the structure adopted by the ammonia molecule and explain using VSEPR rules why this molecule is non-planar.

(b) Give an explanation for the fact that ammonia has higher melting and boiling points than phosphine (PH_3) even though phosphine has a higher relative molecular mass.

(c) Suggest whether the molecule stibine (SbH_3) would have:

i. A pyramidal or planar structure;

ii. A boiling point higher or lower than that of phosphine.

(d) For each of the following examples select the appropriate molecule from the list and give a brief explanation for your answer.

i. Which is the strongest Lewis base: NH_3, PH_3, or SbH_3?

ii. Which is the most stable with respect to decomposition to its constituent elements: NH_3, PH_3, or SbH_3?

Question 3.12

(a) Give the structures of the phosphorus oxides P_4O_6 and P_4O_{10}.

(b) Compare these structures to that of white phosphorus and comment on any similarities.

(c) Would you expect i. nitrogen and ii. arsenic to form oxides with analogous structures?

(d) Give the structures of the phosphorus oxoacids H_3PO_4, H_3PO_3, and H_3PO_2.

(e) What is the oxidation state of the phosphorus atom in each of these oxoacids?

(f) Comment on the fact that: H_3PO_4 can form three types of salts with anions of the formula $H_2PO_4^-$, HPO_4^{2-}, and PO_4^{3-}; H_3PO_3 can form two types of salts with anions of formula $H_2PO_3^-$ and HPO_3^{2-}; while H_3PO_2 forms salts only with the anion $H_2PO_2^-$.

3.7 Group 16

As with group 15, the chemistry of group 16 is dominated by covalent chemistry. The ionization enthalpies of these elements, which lie to the right-hand side of the periodic table are too high to readily allow the formation of cations. However, at the head of the group, some anionic chemistry is observed where ionic oxides (O^{2-}) and sulphides (S^{2-}) are seen. The formation of anions in these cases is typically favoured by the small size of the resultant anion giving rise to a high lattice enthalpy for the product.

Worked example 3.7A

This question concerns the hydrides of group 16 and draws an important distinction between the effects of **intermolecular** hydrogen bonding and **intramolecular** covalent bonding between the central group 16 atom and the hydrogen atoms. The presence of strong intermolecular bonding raises the melting and boiling points of the molecules. In contrast, the intramolecular bonding determines how stable the compounds are with respect to decomposition to their constituent elements. We usually encounter this quantity in its reverse form: the **enthalpy of formation** details the enthalpy change when the material is formed from its constituent elements.

This table gives some data for the group 16 hydrides, H_2O, and H_2Se.

	Melting point/°C	Boiling point/°C	$\Delta_f H^\ominus$/kJ mol^{-1}
H_2O	0	100	−242
H_2Se	−60	−41	+86

(a) Explain why the melting point and boiling point are much higher for H_2O than they are for H_2Se.

(b) Define the term $\Delta_f H^\ominus$ (the standard enthalpy of formation), giving balanced chemical equations to illustrate your definition for the molecules H_2O and H_2Se.

(c) Comment on the fact that $\Delta_f H^\ominus$ for H_2O is negative while that for H_2Se is positive and suggest a reason for this difference based on the bonding within the molecules. State which of the two molecules is more stable.

(d) Would you expect H_2Te to be more, or less, stable than H_2Se, giving a brief explanation for your answer?

Solution

(a) This is an effect of the stronger H-bonding in H_2O caused by the fact that O is much more electronegative than Se. This intermolecular bonding between H_2O molecules increases the boiling and melting points.

(b) The standard enthalpy change per mole of the compound (in its standard state) from its constituent elements in their standard states.

$$H_2(g)+\frac{1}{2}O_2(g)\rightarrow H_2O\,(l)$$

$$H_2(g)+Se\,(s)\rightarrow H_2Se\,(g)$$

(c) This shows that water is stable with respect to oxygen and hydrogen but that H_2Se is unstable with respect to selenium and hydrogen. There are other factors involved, but a principal factor is the strength of the covalent **intramolecular** O—H or Se—H bonding. The O—H bond is much stronger than the Se—H bond due to the relative sizes and energies of the overlapping orbitals. The valence orbitals in oxygen are much closer in energy and size to the 1*s* orbital on hydrogen than the valence orbitals on selenium. This more favourable overlap results in a greater bond strength, which enhances the stability of the molecules.

(d) As Te is lower in the group and hence even larger than Se, you would expect H_2Te to be even less stable because the **intramolecular** Te—H bonding will be even weaker than the Se—H bonding. This is found to be the case, as evidenced by the standard enthalpy of formation of TeH_2: +99.58 kJ mol^{-1}.

Worked example 3.7B

(a) Give the structures of the following group 16 oxoacids:

i. Sulfuric acid (H_2SO_4).

ii. Peroxodisulfuric acid ($H_2S_2O_8$).

iii. Thiosulfuric acid ($H_2S_2O_3$).

iv. Telluric acid (H_6TeO_6).

(b) Explain why peroxodisulfuric acid and telluric acid act as **oxidizing** agents, giving balanced ionic half-equations to illustrate your answer.

(c) Given the following standard electrode potentials:

Equation	$E^{\ominus}/V$
i. $[S_2O_8]^{2-} + 2e^- \rightarrow 2[SO_4]^{2-}$	+2.01
ii. $[Cr_2O_7]^{2-} + 14H^+ + 6e^- \rightarrow 2Cr^{3+} + 7H_2O$	+1.33
iii. $[MnO_4]^- + 8H^+ + 5e^- \rightarrow Mn^{2+} + 4H_2O$	+1.51

Explain whether or not peroxodisulfate ion ($S_2O_8^{2-}$) could be made (at pH = 0) from SO_4^{2-} by oxidation by i. $Cr_2O_7^{2-}$ and/or ii. MnO_4^-. Write overall ionic equations for these redox processes.

Solution

(a)

i. 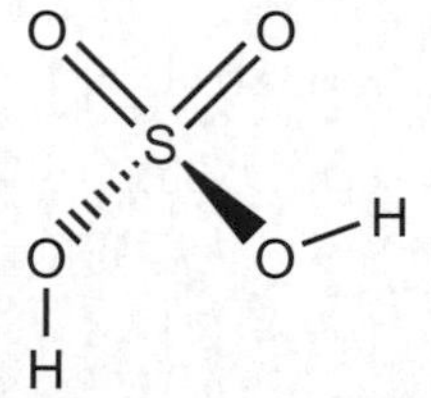

ii. 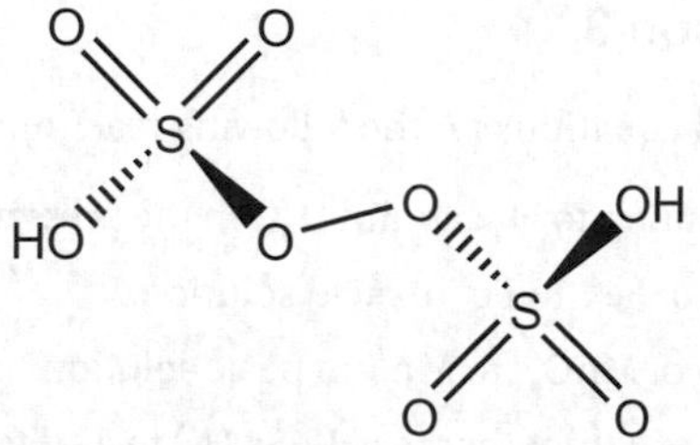

iii. 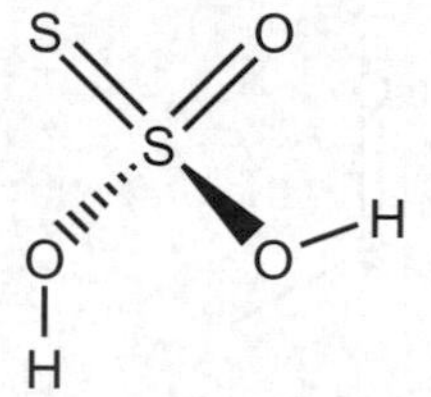

iv. 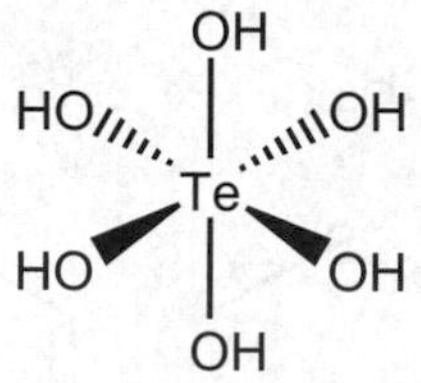

(b) In this question it is necessary to consider what the stated agents will be **reduced** to when they act as oxidizing agents. This will allow us to understand the chemical processes taking place more fully. When peroxodisulfuric acid (or its anion peroxodisulfate—$S_2O_8^{2-}$) acts as an oxidant it is reduced to sulfuric acid (or sulfate—SO_4^{2-}). It is the breaking of the peroxo (O—O) bond that allows this reagent to act as an oxidizing agent. When telluric acid acts as an oxidizing agent, the Te atom (which is in the +6 oxidation state in H_6TeO_6) is reduced to Te (+4) in H_2TeO_3 and it is this process that causes telluric acid to act as an oxidant. The reason that Te(+6) is readily reduced to Te(+4) is the 'inert pair effect': the $5s^2$ pair of electrons in Te are held closely to the nucleus, and so are not readily oxidized. This effect is also discussed in sections 3.3, 3.4, and 3.5. The relevant half-equations are:

$$H_2S_2O_8 + 2H^+ + 2e^- \rightarrow 2H_2SO_4$$

$$H_6TeO_6 + 2H^+ + 2e^- \rightarrow H_2TeO_3 + 3H_2O$$

(c) The overall ionic equations are:

$$[Cr_2O_7]^{2-} + 14H^+ + 6[SO_4]^{2-} \rightarrow 2Cr^{3+} + 3[S_2O_8]^{2-} + 7H_2O$$

$$E^{\ominus} = -0.68\ V$$

$$2[MnO_4]^- + 16H^+ + 10[SO_4]^{2-} \rightarrow 2Mn^{2+} + 5[S_2O_8]^{2-} + 8H_2O$$

$$E^{\ominus} = -0.50\,V$$

In both cases there is an overall negative standard cell potential. This indicates that the free energy change for the process would be positive (unfavourable) from:

$$\Delta G^{\ominus} = -zFE^{\ominus}$$

where F is the Faraday constant.

As such it is shown that neither $Cr_2O_7^{2-}$ nor MnO_4^- can be used to oxidize SO_4^{2-} to form $S_2O_8^{2-}$.

Question 3.13

(a) Which bond is stronger S=S or O=O?

(b) Does gaseous SO_2 have a linear or a bent structure?

(c) Can $H_2S_2O_3$ be isolated as a free acid? Can its salts—those of the ion $S_2O_3^{2-}$—b e isolated?

(d) Is the bond distance in O_2^{2-} longer or shorter than that in O_2?

(e) Can O_2^+ be formed by oxidation of O_2 by PtF_6?

Question 3.14

Give balanced equations for the following reactions of hydrogen peroxide (H_2O_2):

(a) Decomposition to O_2 gas and H_2O in the presence of a catalyst such as MnO_2.

(b) Oxidation of Fe^{2+} to Fe^{3+} in acid solution.

(c) Reduction of MnO_4^- to Mn^{2+} in basic solution.

(d) Oxidation of a thioether (e.g. $PhSCH_3$) to a sulfoxide (e.g. $PhS(O)CH_3$).

(i) $PhSCH_3$ (ii) $PhS(O)CH_3$

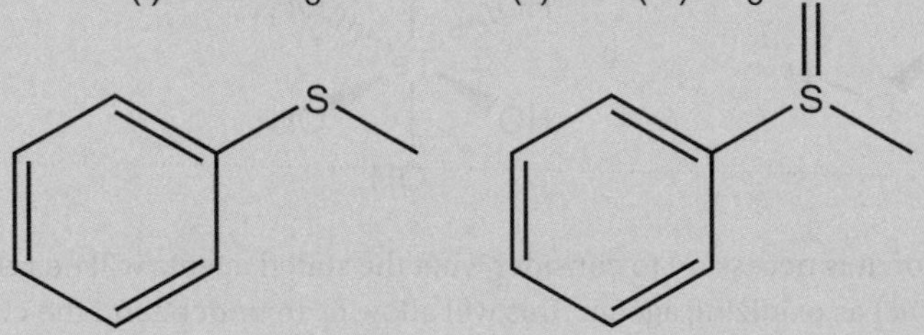

(e) Reduction of sodium hypochlorite to sodium chloride in basic solution.

3.8 Group 17

The elements of group 17, also known as the **halogens**, exhibit a wide variety of covalent and ionic chemistry. The halogen elements are highly reactive, and, as a result, halogen compounds of almost every element are known.

Due to their position at the right-hand side of the periodic table, the group 17 elements have high ionization enthalpies, and as such, cationic chemistry is not favoured. However, highly exothermic electron affinities ($\Delta_{EA}H^{\ominus}$) are observed for all members of the group and so the group 17 elements readily form anions with the –1 oxidation state.

As well as anionic behaviour, a wide variety of covalent halogen compounds and complex ions are known, including **interhalogen compounds** and **polyhalide ions**, which feature two or more different halogen elements bonded together.

Worked example 3.8A

(a) In the 'iodine clock' reaction I_2 is formed by the reaction of iodide (I^-) and iodate (IO_3^-) ions in the presence of dilute acid. Give balanced half-equations and an overall balanced ionic equation for this reaction.

(b) Iodine monofluoride (IF) disproportionates into IF_5 and I_2.

i. Explain what is meant by the term **disproportionation**.

ii. Give a balanced equation for the reaction.

iii. Given that the standard enthalpies of formation ($\Delta_f H^\ominus$) of IF, IF_5, and I_2 are −95, −840, and 0 kJ mol^{-1} respectively, calculate $\Delta_f H^\ominus$ for the disproportionation reaction that you have given in ii.

(c) Use these data to calculate the bond enthalpy of the I_2 molecule (assuming that the I—F bond enthalpy in IF is the same as that in IF_5), commenting on the value you obtain relative to the bond enthalpy of the Cl_2 molecule of 242 kJ mol^{-1}.

Solution

(a) The 'iodine clock' reaction consists of a complex series of reactions. In this particular step elemental I_2 is formed by the oxidation of I^- by IO_3^- in which the iodine atom is in the +5 oxidation state. The way to approach balancing a redox ionic equation of this type is first of all to write balanced ionic equations for the reduction and oxidation steps:

Reduction: $IO_3^-(aq)+6H^+(aq)+5e^- \rightarrow \frac{1}{2}I_2(aq)+3H_2O\,(l)$

Oxidation: $I^-(aq)\rightarrow \frac{1}{2}I_2(aq)+e^-$

Note that in the reduction step the equation must be balanced chemically with respect to oxygen and hydrogen atoms. As the reaction occurs in acidic solution this is done by adding the appropriate number of water molecules to the right-hand side (three in order to balance the three oxygen atoms from IO_3^- in three molecules of H_2O) and then balancing these by adding the appropriate number of H^+ ions (six) to the left-hand side.

To produce a balanced ionic equation it is necessary to add these two equations together, having multiplied the oxidation reaction by five so that the total number of electrons on each side of the equation cancels out.

This gives us:

$$IO_3^-(aq)+6H^+(aq)+5I^-(aq) \rightarrow 3I_2(aq)+3H_2O\,(l)$$

(b)

i. A disproportionation reaction is one in which a species in one oxidation state spontaneously converts to species in higher and lower oxidation states.

ii. The balanced equation for this reaction is:

$$5IF \rightarrow 2I_2 + IF_5$$

Note that in this reaction iodine is in the +1 oxidation state in IF disproportionates into iodine in the +5 (in IF_5) and 0 (in I_2) oxidation states.

iii. In order to calculate the overall standard enthalpy of reaction from the standard enthalpies of formation one must subtract the sum of the standard enthalpy of formation of the products from the sum of the standard enthalpies of formation of the reactants.

$$\Delta_r H^\ominus = \Sigma\Delta_f H^\ominus(\text{products}) - \Sigma\Delta_f H^\ominus(\text{reactants})$$

This gives us:

$$\Delta_r H^\ominus = \{\Delta_f H^\ominus(IF_5) + 2\,\Delta_f H^\ominus(I_2)\} - \{\Delta_f H^\ominus(IF)\}$$

$$\Delta_r H^\ominus = \{-840+0\} - \{(5\times -95)\} = \{-840\} + \{475\} \quad \text{kJ mol}^{-1}$$

$$\Delta_r H^\ominus = -365 \text{ kJ mol}^{-1}$$

(c) We can also calculate the standard enthalpy of reaction in terms of sum of the bond dissociation enthalpies of the bonds **formed** (exothermic) minus the sum of the bond dissociation enthalpies of the bonds **broken** (endothermic).

$$\Delta_r H^\ominus = \Sigma D_e(\text{bonds formed}) - \Sigma D_e(\text{bonds broken})$$

This chemistry gives good examples of **interhalogen** compounds. These are covalent compounds that include more than one type of halogen atom. In this case both IF and IF_5 are interhalogen compounds.

It can be seen that there are the same number of I—F bonds in reactants and products (five in each) so the only new bonds created are two I—I bonds. If we assume that the I—F bonds in IF and IF_5 have the same bond dissociation enthalpy, then each of the I—I bonds must have a bond dissociation enthalpy of 365/2 = 183 kJ mol^{-1}. As expected, this is lower than the bond dissociation enthalpy of Cl—Cl in Cl_2, as orbital overlap is weaker.

Worked example 3.8B

(a) Give a general chemical equation to show how Ag^+ ions react with halide ions in aqueous solution.

(b) Give a chemical equation to show how the halide AgCl dissolves in excess aqueous ammonia solution.

(c) Explain how the reactions you give in parts (a) and (b) might be used to distinguish between the halide ions Cl^-, Br^-, and I^-.

(d) Give a chemical equation to show how Cu^{2+} ions react with I^- ions in aqueous solution.

(e) Give a chemical equation to show how the solid product of this reaction redissolves in excess aqueous ammonia solution.

(f) In an iodine–thiosulfate titration, thiosulfate is oxidized to tetrathionate ($S_4O_6^{2-}$).

i. Give a balanced ionic equation for the reaction.

ii. Give the average oxidation state of sulfur in thiosulfate and tetrathionate.

iii. Explain, using ionic equations where appropriate, how the reactions in parts (c) and (e) could be used to determine the concentration of copper in a copper(II) compound e.g. $CuSO_4.5H_2O$.

Solution

(a) This is a classic reaction of the halide ion with Ag^+ to form the AgX (X = halide) solid.

$$Ag^+(aq) + X^-(aq) \rightarrow AgX(s)$$

(b) The silver halide solids dissolve to a greater or lesser extent in aqueous ammonia solution according to the equilibrium:

$$AgX(s) + 2NH_3(aq) \rightleftharpoons [Ag(NH_3)_2]^+(aq) + X^-(aq)$$

(c) Reactions of parts (a) and (b) can be used together to identify the halide ions in terms of i. colour of the precipitate and ii. solubility in concentrated ammonia solution. AgCl is white, AgBr is off-white (cream-coloured), and AgI is yellow. These colour changes can be accounted for by **charge-transfer bands**. Ag^+ has a filled $4d^{10}$ shell of electrons and as such cannot display *d-d* transitions. Thus the colour must be caused by charge transfer from the halide ion to the silver. This process becomes energetically more favourable as the group is descended as the halide ions become more polarizable and more readily oxidized.

The colours we observe result from light of certain wavelengths being absorbed and others transmitted or reflected by a material. For the AgCl precipitate, the absorbance band lies in the UV part of the spectrum, hence all visible light is reflected and hence it appears white to us. For AgI the absorbance band moves to the visible region, so some visible light is absorbed. The particular wavelengths absorbed mean that we perceive a yellow colour for the precipitate due to the wavelengths reflected.

The solubility of the halide salt in concentrated ammonia solution decreases as the group is descended: AgCl is soluble, AgBr is slightly soluble, and AgI is insoluble. This is in contrast to the order expected on the basis of the lattice enthalpies of the salts, which will decrease down the group as the anions become larger. Thus the decreased solubility is driven by the decreasing enthalpies of hydration of the halide ions as the group is descended.

(d) In this case the Cu^{2+} acts as an **oxidant**. Unlike silver, copper can exist in +1 and +2 oxidation states. Thus, in this reaction it is reduced to copper in the +1 oxidation state in the compound CuI which is insoluble and precipitates. The I^- ions are oxidized to I_2:

$$Cu^{2+}(aq)+2I^-(aq)\rightarrow CuI(s)+\tfrac{1}{2}I_2(aq)$$

(e) When CuI re-dissolves in aqueous ammonia solution, the Cu is **oxidized** to Cu^{2+}:

$$2CuI(s)+I_2(aq)+4NH_3(aq)+2H_2O\rightarrow 2[Cu(NH_3)_4(H_2O)_2]^{2+}(aq)+4I^-(aq)$$

It may be asked why is CuI oxidized to Cu^{2+} in aqueous ammonia solution when Cu^{2+} is reduced by I^- in water? The answer is that the Cu in the +2 oxidation state is stabilized by the presence of NH_3 ligands in the complex ion $[Cu(NH_3)_4(H_2O)_2]^{2+}$ making its formation more thermodynamically favourable.

(f)

i. The balanced equation is:

$$I_2(aq)+\ 2S_2O_3^{2-}(aq)\rightarrow 2I^-(aq)+S_4O_6^{2-}(aq)$$

ii. To calculate the oxidation states of the sulfur it is necessary to consider the number of O atoms in the ions, each of which may be assigned a –2 oxidation state.
Thus in $S_2O_3^{2-}$ we have a total of three O atoms, each with a formal –2 oxidation state and an overall 2– charge on the ion. Thus each S atom must have an oxidation state of $4/2=2$, so each S atom has an oxidation state of +2.
In $S_4O_6^{2-}$ we have a total of six O atoms each with a formal –2 oxidation state and an overall 2– charge on the ion. So each S atom has a formal $+10/4=+2.5$ oxidation state. Clearly the S atoms have been **oxidized** as the I_2 is **reduced** to I^-.

iii. By combining the equations in parts (d) and (f) it can be seen that I_2 may be produced stoichiometrically by adding Cu^{2+} to an excess of I^-. The I_2 may then be titrated against standardized thiosulfate ($S_2O_3^{2-}$) solution. By calculating the amount in moles of $S_2O_3^{2-}$ used in the titration it is possible to calculate the amount in moles and hence the mass of copper in the original sample.

At first sight it may appear rather odd that here we have a fractional oxidation state of +2.5. Clearly we cannot divide an electron in half, so what is occurring here? We must consider the structure of the $S_4O_6^{2-}$ ion:

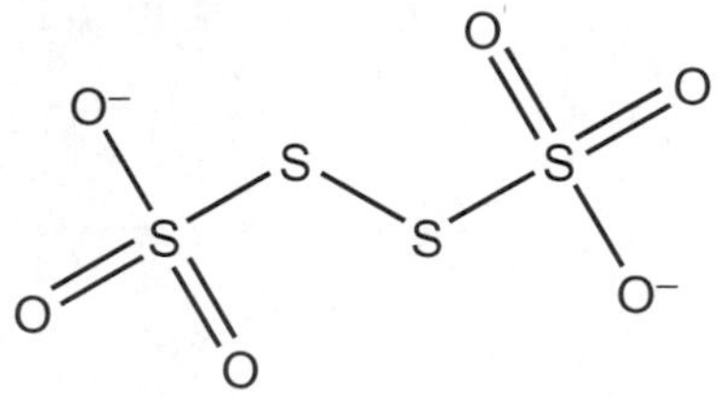

There are two types of S atom within this ion. We consider bonds between atoms of the same element to have no effect on their oxidation state, so the central S atoms, which form bonds **only** to other sulfur atoms, are in oxidation state 0. The two non-central S atoms are each linked to three O atoms, via two double bonds and one single bond. Therefore, because oxygen is more electronegative than sulfur, these atoms each have a formal oxidation state of +5. Thus the overall **average** oxidation state of the four S atoms in this ion is +2.5.

Question 3.15

(a) Explain why fluorine (F_2) is extracted by electrolysis of molten fluorides whereas bromine and iodine may be produced chemically by oxidation of bromides and iodides respectively.

(b) Explain why iodine has a purple colour in non-coordinating solvents such as hexane but has a brownish colour in coordinating solvents (those with lone pairs of electrons) such as tetrahydrofuran (THF) (C_4H_8O).

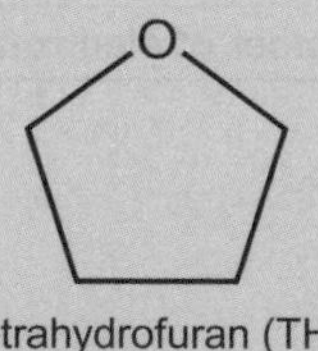

tetrahydrofuran (THF)

(c) Suggest an explanation for the fact that the halides of non-metals and metals in high oxidation states tend to be covalent while those of metals in low oxidation states tend to be ionic.

(d) Explain why the acidity of the chlorine oxoacids $HOClO_n$ ($n=0,1,2,3$) increases as n (the number of oxygen atoms) increases.

Question 3.16

(a) An iodimetric titration was carried out to determine the amount of copper in an unknown copper(II) salt. A 0.1000 g sample of the salt was weighed out and was reacted with potassium iodide solution. The liberated iodine was titrated against 0.0120 mol dm^{-3} $Na_2S_2O_3$ solution. It was found that 32.50 cm^3 of the $Na_2S_2O_3$ solution was need to react with all of the iodine. Calculate the percentage of copper in the unknown copper salt.

(b) It is suggested that the unknown copper salt was, in fact, hydrated copper(II) sulfate. From your calculation in part (a) estimate the purity of the sample stating any assumptions that you have made.

3.9 Group 18

The monatomic gases that make up group 18 were originally known as the **inert gases** due to their unreactive nature. Their position at the right-most edge of the periodic table reflects both their 'full shell' electronic configurations and high ionization enthalpies. Whilst, in general, the elements of group 18 show much more limited chemistry in comparison to the rest of the periodic table, a number of interesting compounds are now known, particularly for the elements lower in the group, such as krypton and xenon.

Worked example 3.9A

Xenon forms the fluorides XeF_2, XeF_4, and XeF_6.

(a) Sketch the structures of these fluorides and discuss **briefly** whether VSEPR (valence shell electron pair repulsion) theory accurately predicts these structures.

(b) XeF_6 is readily hydrolysed to give XeO_3. Give a balanced equation for this reaction. Also give balanced equations for the reaction of XeO_3 with OH^- ions to give $[HXeO_4]^-$ and of $[HXeO_4]^-$ with OH^- ions to give $[XeO_6]^{4-}$, alongside Xe and O_2 gases. What do you think would be the specific safety concerns in carrying out this chemistry?

Solution

(a) The AXE method (which is outlined in Chapter 2) is used to determine the VSEPR predicted shapes of the molecules XeF_2, XeF_4, and XeF_6.

XeF_2	Number of electrons
Valence electrons on Xe	8
Electrons from each F	1×2
0 charge	0
Zero *n*th order bonds	0
Total number of electrons	10

There are ten electrons predicted in total, meaning five pairs of electrons. With two fluorine atoms bonded to the central xenon atom, this corresponds to two bonding pairs and three lone pairs (AX_2E_3) and therefore VSEPR predicts a linear molecule:

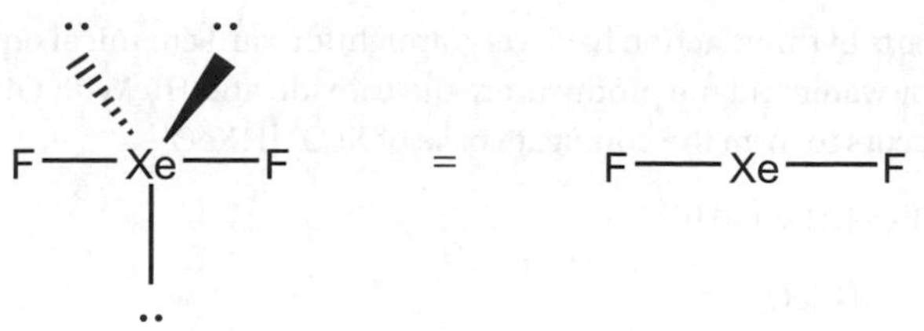

Note that in this case the structure is linear. Lone pair–lone pair repulsion is greater than lone pair–bonding pair repulsion. By adopting a linear structure the three lone pairs are at 120° to each other.

This structure agrees well with experiment.

XeF_4	Number of electrons
Valence electrons on Xe	8
Electrons from each F	1 × 4
0 charge	0
Zero *n*th order bonds	0
Total number of electrons	12

Twelve electrons are predicted in total, meaning six pairs of electrons. With four fluorine atoms bonded to the central xenon atom, this corresponds to four bonding pairs and two lone pairs (AX_4E_2) and therefore VSEPR predicts a square-planar molecule:

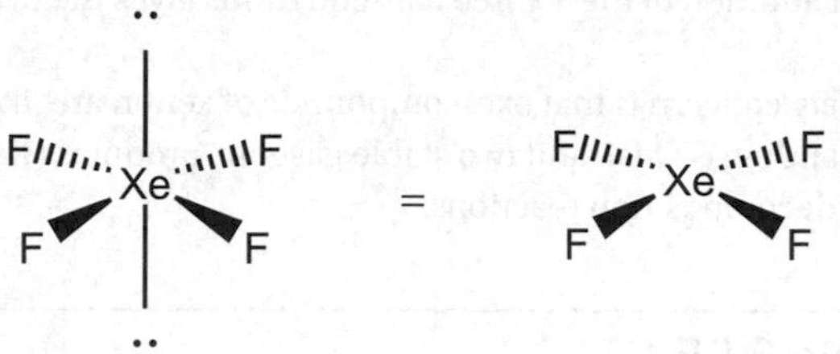

This structure agrees well with experiment.

XeF_6	Number of electrons
Valence electrons on Xe	8
Electrons from each F	1 × 6
0 charge	0
Zero *n*th order bonds	0
Total number of electrons	14

Fourteen electrons are predicted in total, meaning seven pairs of electrons. With six fluorine atoms bonded to the central xenon atom, this corresponds to six bonding pairs and one lone pair (AX_6E_1) and therefore VSEPR predicts a pentagonal-based pyramidal molecule:

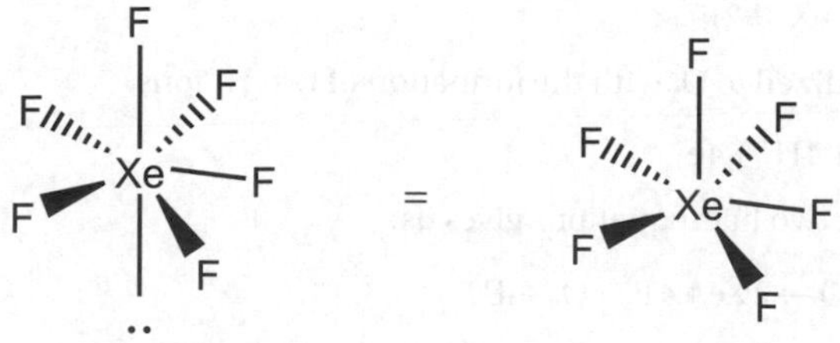

This structure does not agree with experiment, though determining the precise shape of XeF_6 is not easy. The structure adopted by XeF_6 is based on an octahedron; according to Konrad Seppelt, an expert on noble gas and fluorine chemistry, 'the structure is best described in terms of a mobile electron pair that moves over the faces and edges of the octahedron and thus distorts it in a dynamic manner' (Seppelt 1979).

(b) The first two parts of this reaction have very straightforward chemical equations. XeF_6 is hydrolysed by water and the products are the trioxide and HF. With OH^- an acid–base equilibrium occurs to form the conjugate base of XeO_3 $[HXeO_4]^-$.

$$XeF_6 + 3H_2O \rightarrow XeO_3 + 6HF$$

$$XeO_3 + OH^- \rightarrow [HXeO_4]^-$$

The last part of this reaction is much more complex and two gaseous products are formed namely Xe and O_2.

$$2[HXeO_4]^- + 2OH^- \rightarrow [XeO_6]^{4-} + Xe + O_2 + 2H_2O$$

Disproportionation is a redox process in which an element in a particular oxidation state is simultaneously oxidized and reduced. In this case Xe(+6) is oxidized to Xe(+8) and reduced to Xe(0).

Clearly redox chemistry and **disproportionation** is taking place here in that some of the Xe(+6) atoms in $[HXeO_4]^-$ have been reduced to xenon gas (0 oxidation state) while others have been oxidized to Xe(+8) in $[XeO_6]^{4-}$. At the same time OH^- has been oxidized to O_2.

This reaction is best considered in terms of three half-equations:

$$[HXeO_4]^- + 5OH^- \rightarrow [XeO_6]^{4-} + 2e^- + 3H_2O$$

$$[HXeO_4]^- + 6e^- + 3H_2O \rightarrow Xe + 7OH^-$$

$$4OH^- \rightarrow 2H_2O + O_2 + 4e^-$$

It can be seen that addition of these three half-equations gives the overall balanced ionic equation.

The specific safety concern is that oxo-compounds of xenon are likely to be highly explosive. Their bonding is weak and two stable gaseous products (Xe and O_2) are very likely products of decomposition reactions.

Worked example 3.9B

(a) XeF_2 oxidizes H_2O to O_2, itself being reduced to Xe. Give balanced half-equations for the oxidation and reduction reactions and hence give an overall balanced equation for the reaction.

(b) Sketch diagrams to show how the Xe 5*p* and F 2*p* orbitals can overlap in XeF_2 to give σ-bonding, non-bonding, and antibonding orbitals. Hence give a simple molecular orbital diagram which accounts for the stability of XeF_2. Use MO theory to explain why XeF_2 has weaker Xe—F bonds than does $[XeF]^+$.

Solution

(a) Xenon difluoride is reduced to xenon with the formation of two fluoride ions; its oxidation state falls from +2 to 0.

$$XeF_2 + 2e^- \rightarrow Xe + 2F^-$$

H_2O is also **oxidized** to O_2 with the formation of two H^+ ions.

$$2H_2O \rightarrow O_2 + 4H^+ + 4e^-$$

Combing these two half-equations gives us:

$$2XeF_2 + 2H_2O \rightarrow 2Xe + 4F^- + O_2 + 4H^+$$

(b) A diagram to show how the Xe 5*p* and F 2*p* orbitals can overlap in XeF_2 to give σ-bonding, non-bonding, and antibonding orbitals is given in Figure 3.14. A molecular orbital diagram which accounts for the stability of XeF_2 is shown in Figure 3.15.

A total of four electrons bond together the F–Xe–F unit (an electron pair from Xe(5*p*) and one electron in each of the F(2*p*) orbitals) by way of a three-centre, four-electron bond.

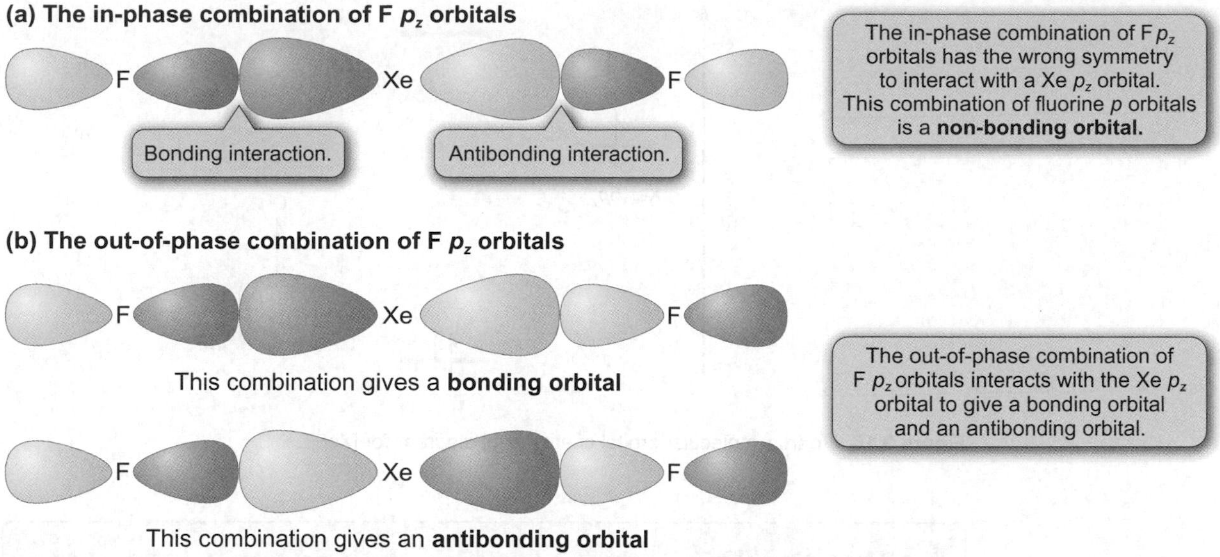

(a) The in-phase combination of F p_z orbitals

Bonding interaction.

Antibonding interaction.

> The in-phase combination of F p_z orbitals has the wrong symmetry to interact with a Xe p_z orbital. This combination of fluorine p orbitals is a **non-bonding orbital.**

(b) The out-of-phase combination of F p_z orbitals

This combination gives a **bonding orbital**

> The out-of-phase combination of F p_z orbitals interacts with the Xe p_z orbital to give a bonding orbital and an antibonding orbital.

This combination gives an **antibonding orbital**

Figure 3.14 The interactions between the combinations of F($2p_z$) orbitals and Xe($5p_z$) orbitals gives rise to bonding, non-bonding, and antibonding molecular orbitals.

These electrons fill the bonding and non-bonding molecular orbitals meaning that this molecule has two bonding electrons distributed across three centres. The remaining electrons on the Xe and F atoms are non-bonding and are present in the Xe 5*s* and $5p_{x,y}$ and F 2*s* and $2p_{x,y}$ orbitals. The bonds in $[XeF]^+$ contain a net of two electrons in a bonding molecular orbital (it is isoelectronic to the interhalogen compound IF); see Figure 3.16. Thus XeF_2 has two bonding electrons holding together three atoms, whereas $[XeF]^+$ has two bonding electrons holding together two atoms and so the Xe—F bonds in $[XeF]^+$ are stronger.

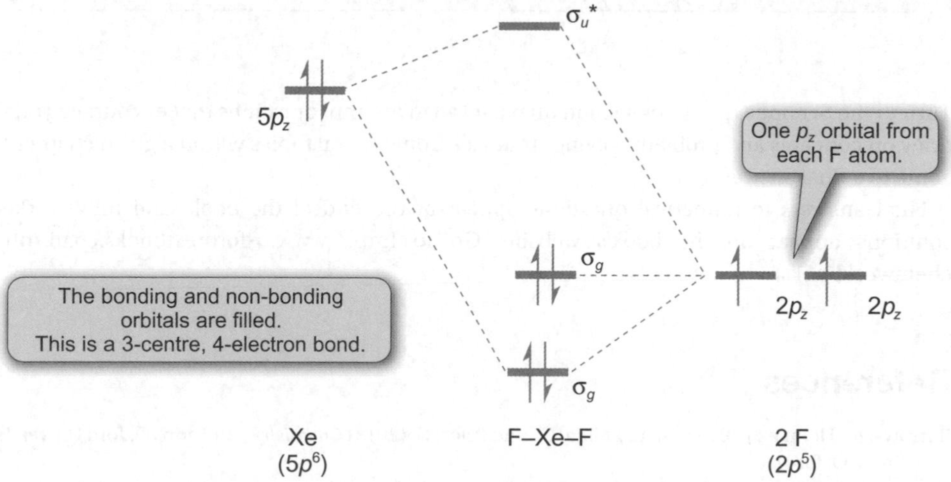

One p_z orbital from each F atom.

The bonding and non-bonding orbitals are filled. This is a 3-centre, 4-electron bond.

σ_u^*

$5p_z$

σ_g

$2p_z$ $2p_z$

σ_g

Xe
($5p^6$)

F–Xe–F

2 F
($2p^5$)

Figure 3.15 A partial molecular orbital energy level diagram for XeF_2.

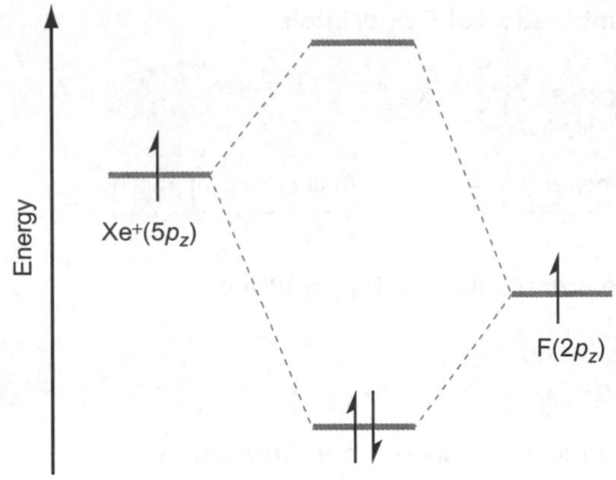

Figure 3.16 A partial molecular orbital energy level diagram for $[XeF]^+$.

Question 3.17

Suggest products for the following reactions:

(a) CsF and XeF_6.

(b) XeF_6 and SbF_5.

(c) KrF_2 and H_2O.

(d) XeF_2 and Ir.

(e) XeF_2 and S.

Question 3.18

Comment on why the chemistry of the xenon fluorides is much more understood than that of: i. the xenon chlorides; ii. the krypton fluorides and iii. the radon fluorides. Look up some examples of chemical compounds in each of these categories and comment on the examples that you have found.

Turn to the Synoptic questions section on page 148 to attempt questions that encourage you to draw on concepts and problem-solving strategies from several topics within a given chapter to come to a final answer.

Final answers to numerical questions appear at the end of the book, and fully worked solutions appear on the book's website. Go to http://www.oxfordtextbooks.co.uk/orc/chemworkbooks/.

References

Burrows, A., Holman, J., Parsons, A., Pilling, G., and Price, G. (2013) *Chemistry*^{3}, 2nd edn (Oxford University Press, Oxford).

Seppelt, K. (1979) 'Recent developments in the chemistry of some electronegative elements', *Acc. Chem. Res.*, **12** (6), 211–216.

4 Solids

4.1 The packing of spheres to give solid structures

Close-packed structures

Atoms can be considered as small hard spheres. Extended structures such as those of metallic elements can be considered to be made up of atoms packed together in a regular array. When atoms or spheres are packed together there are two possible types of arrangements for the first layer of atoms. The spheres can either sit next to each other such that each sphere touches four others in the same layer (Figure. 4.1a), or the atoms can be shifted horizontally or vertically by half a radius from this arrangement such that they pack more closely to each other and take up less space (Figure 4.1b).

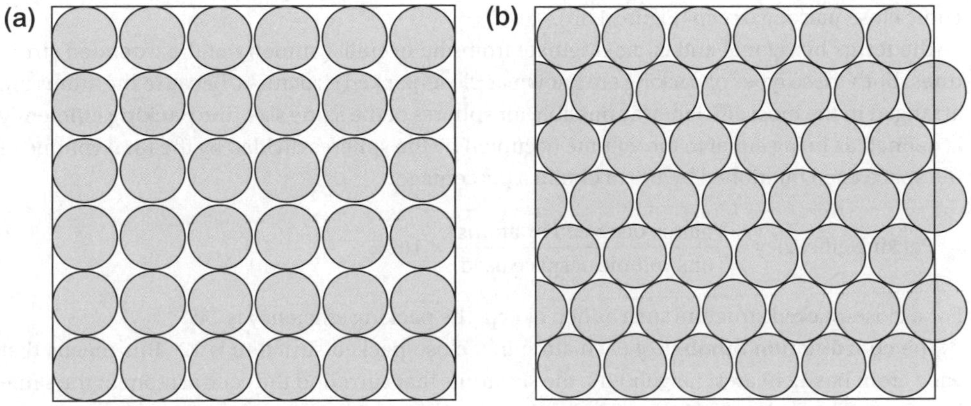

Figure 4.1 (a) A square layer of spheres. (b) A close-packed layer of spheres.

This is a more efficient way of packing spheres; in this arrangement each atom in the layer touches six other atoms in the same layer.

To build up a three-dimensional array of spheres, a second layer of atoms must be placed on top of the first. The atoms in this second layer sit in half the depressions or holes left by the first layer in Figure 4.1b—these are alternate holes (Figure 4.2).

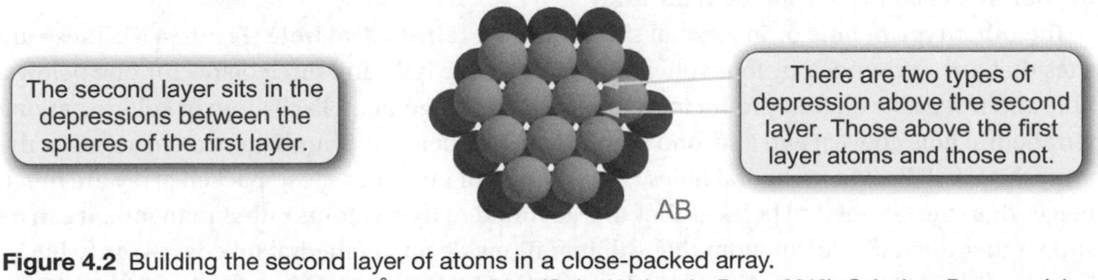

The second layer sits in the depressions between the spheres of the first layer.
There are two types of depression above the second layer. Those above the first layer atoms and those not.
AB

Figure 4.2 Building the second layer of atoms in a close-packed array.
Adapted from Burrows et al., *Chemistry*[3] second edition (Oxford University Press, 2013). © Andrew Burrows, John Holman, Andrew Parsons, Gwen Pilling, and Gareth Price 2013.

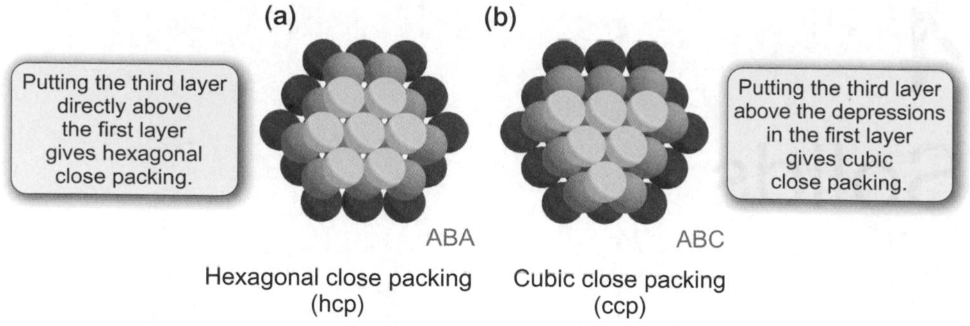
(a)
(b)
Putting the third layer directly above the first layer gives hexagonal close packing.
Putting the third layer above the depressions in the first layer gives cubic close packing.
ABA
ABC
Hexagonal close packing (hcp)
Cubic close packing (ccp)

Figure 4.3 Building the third layer of atoms in a close-packed array. (a) The ABA . . . arrangement of a hexagonal close-packed array (hcp). (b) The ABC . . . arrangement of a cubic close-packed array (ccp). Adapted from Burrows et al., *Chemistry*[3] second edition (Oxford University Press, 2013). © Andrew Burrows, John Holman, Andrew Parsons, Gwen Pilling, and Gareth Price 2013.

When a third layer is added there are two possibilities for the placement of these atoms. They can either sit in the holes which place them directly over the atoms in the first layer. If we say that the first layer is layer A and the second layer is layer B then the third layer will have the same arrangement as layer A and so we get an **ABAB . . .** repeating sequence (Figure 4.3a).

➲ A cubic close-packed arrangement of spheres is also referred to as a face-centred cubic (**fcc**) arrangement because atoms sit at the corners and centres of the faces of a cube

This type of arrangement is **h**exagonal **c**lose **p**acking or **hcp**. Alternatively we could place the third layer atoms over the holes that are still unoccupied in the first layer. The third layer thus has atoms in different positions from either layer A or layer B and may be designated by the letter C. In this way the repeating sequence is **ABCABC . . . .** This type of arrangement is called **c**ubic **c**lose **p**acking or **ccp** (Figure 4.3b).

The terms hexagonal and cubic originate from the overall symmetry of the extended structures. Both these types of packing are known as close-packed structures, because the atoms are arranged in the most efficient way possible for spheres of the same size. The **packing efficiency** is defined as being equal to the volume occupied by the spheres divided by the total volume of the space used multiplied by 100 to obtain a percentage.

$$\text{Packing efficiency} = \frac{\text{Volume occupied by atoms}}{\text{Total volume of space used}} \times 100\%$$

For a close-packed structure such as hcp or ccp, the packing efficiency is 74%.

The **coordination number** of each atom in a close-packed structure is 12. This means that each atom has 12 nearest neighbours: the six atoms that surround the central atom in the same layer (as we have already discussed) plus three atoms in the layer above and three atoms in the layer below (Figure 4.4).

Even though the atoms are in close-packed arrays, there is still some empty space in the structures. This space is in the form of small holes in between the atoms called **interstitial sites**. These holes are of two different types. One type is called an **octahedral hole** as shown in Figure 4.5. Octahedral holes are holes which are surrounded by six atoms or spheres. The centres of the six spheres lie at the corners of an octahedron. Three of the atoms are in the layer above the hole and three are in the layer below. The number of octahedral holes is the same as the number of atoms or spheres in the array.

The other type of hole or interstitial site is called a **tetrahedral hole** (Figure 4.5). These are sites that are surrounded by four spheres, one above the hole and three below (or one below a hole and three above), thus producing a tetrahedral arrangement. Each atom or sphere has one tetrahedral hole above it and a second tetrahedral hole below it. Thus if there are n atoms in the array there will be $2n$ tetrahedral holes. The octahedral holes in a close-packed array are much bigger than the tetrahedral holes as they are surrounded by six atoms rather than four. It can be shown that the radius of an atom that will **just** fit inside an octahedral hole is $0.414r$ (where r is the radius of the spheres making up the array) and the radius of an atom that will **just** fit in

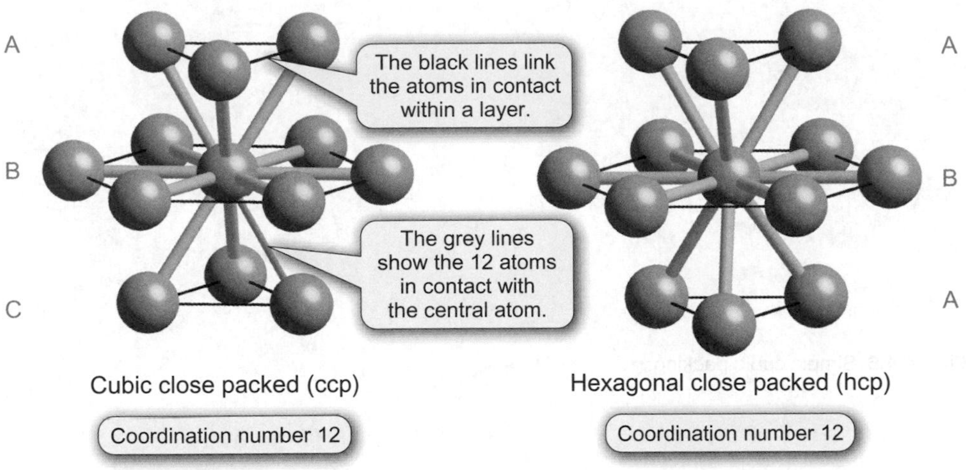

Figure 4.4 The coordination number of an atom in a close-packed array is 12.
Reproduced from Burrows et al., *Chemistry*[3] second edition (Oxford University Press, 2013). © Andrew Burrows, John Holman, Andrew Parsons, Gwen Pilling, and Gareth Price 2013.

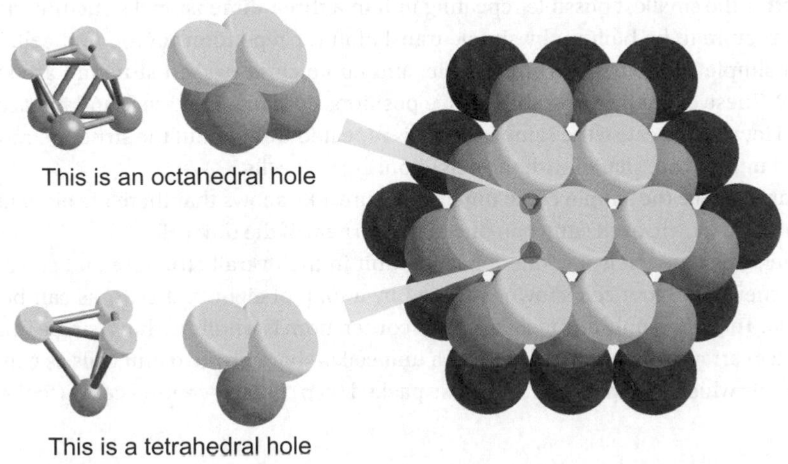

Figure 4.5 Octahedral and tetrahedral holes in a close-packed array.
Reproduced from Burrows et al., *Chemistry*[3] second edition (Oxford University Press, 2013). © Andrew Burrows, John Holman, Andrew Parsons, Gwen Pilling, and Gareth Price 2013.

the tetrahedral hole is $0.225r$. Thus the bigger the atoms the bigger the holes, but the packing efficiency remains the same.

Body-centred cubic and primitive structures

The simplest type of packing is the **primitive** or **simple cubic structure**. The spheres are placed in a square array as shown in Figure 4.6, and the second layer is placed directly above the spheres in the first layer and repeating layers are the same so the arrangement is **AAAA**. Each atom is on the corner of a cube and the coordination number is six. The simple cubic structure is the least efficient way of packing spheres, with a packing efficiency of 52.4%.

The **b**ody-**c**entred **c**ubic structure (**bcc**) is based upon the simple cubic structure but an additional atom has been placed in the centre of each cubic array of atoms (Figure 4.7). This causes the atoms to be forced apart such that the atoms are no longer in contact along the side of the cubic array. Each atom in a body-centred cubic array is eight coordinate and the packing

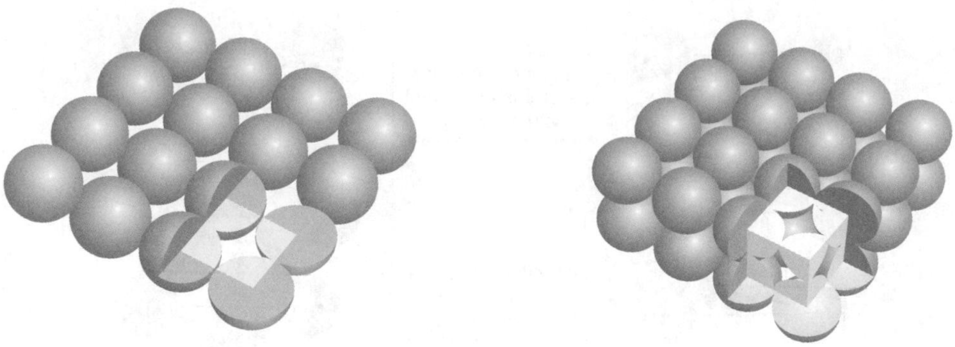

Figure 4.6 Simple cubic packing.

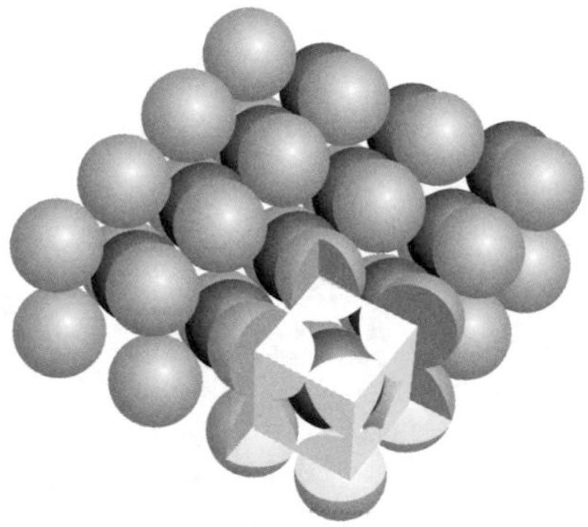

Figure 4.7 Body-centred cubic packing.

➔ Face-centred cubic (fcc) is an alternative name for a cubic close-packed (ccp) arrangement of atoms.

efficiency is 68%. While this is a more efficient packing arrangement than that of primitive cubic structures, it is still less efficient than a close-packed array.

Unit cells

A **unit cell** is the smallest possible repeating unit in a three-dimensional structure that allows the full structure to be built up by simple translation (or repetition) of the unit cell. The unit cells of a simple cubic, body-centred cubic, and cubic close-packed structure are shown in Figure 4.8. These diagrams show the relative positions of atoms that are found in each type of unit cell. However, because the same unit cell is repeated throughout the structure most of the atoms in a unit cell are shared with the neighbouring unit cells.

For example take the simple cubic unit cell. Figure 4.8c shows that there are eight atoms involved in this unit cell, each sitting on one of the corners of the unit cell.

However, this unit cell is just one repeating unit in the overall structure and so each atom on the corner of the unit cell shown is shared by a total of eight unit cells, as can be seen in Figure 4.9a. Therefore, only one eighth of each corner atom is wholly within a single unit cell. In a ccp (or fcc) array any atom on the face of a unit cell is shared by two unit cells as can be seen in Figure 4.9c which represents a cubic close packed (ccp) or face-centred cubic (fcc) array.

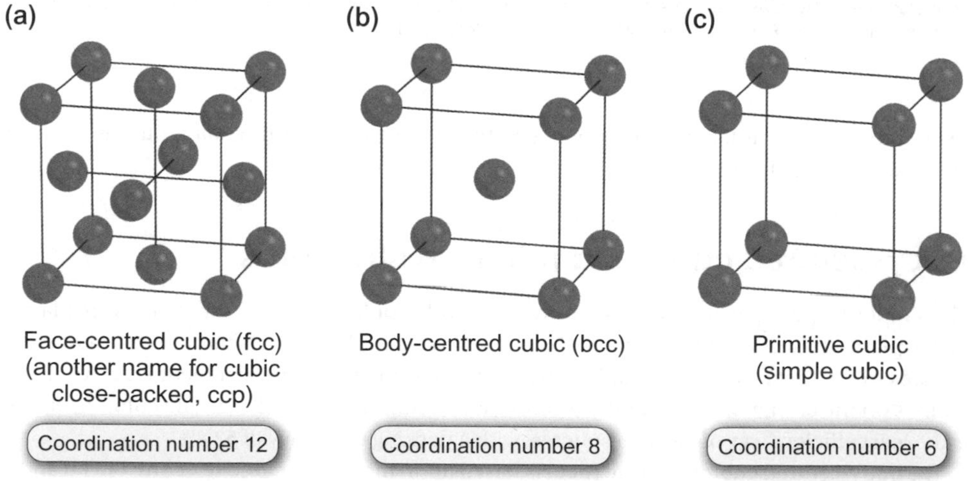

Figure 4.8 Units cells of cubic lattices.
Reproduced from Burrows et al., *Chemistry*^{3} second edition (Oxford University Press, 2013). © Andrew Burrows, John Holman, Andrew Parsons, Gwen Pilling, and Gareth Price 2013.

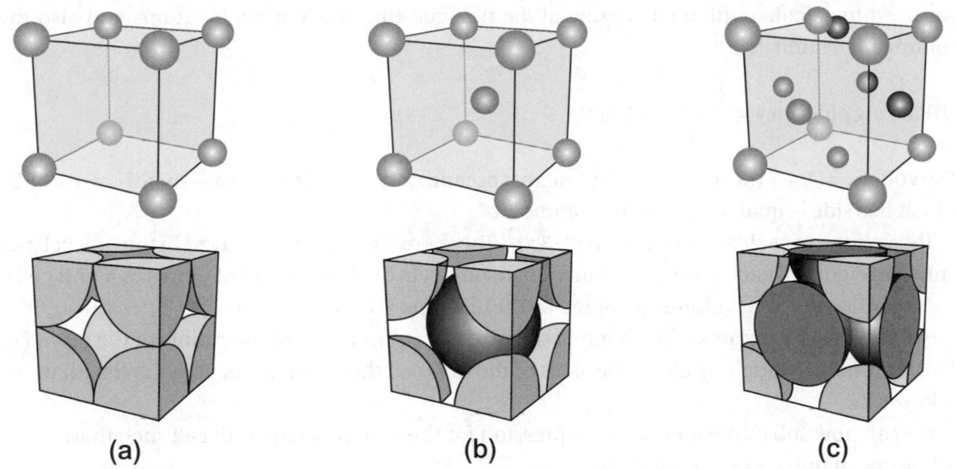

Figure 4.9 Unit cells of cubic lattices: (a) a corner atom in a simple cubic array; (b) a centre atom in a body-centred cubic array; and (c) a centre face atom in a face-centred cubic array.

Therefore, only one half of an atom on a face is present in a single unit cell. Only an atom in the centre of a unit cell, such as in Figure 4.9b which represents a body-centred cubic (bcc) structure, is wholly within a single unit cell.

Worked example 4.1A

Calculate the packing efficiency in a simple cubic structure.

Solution

First of all sketch a unit cell of a simple cubic structure as in Figure 4.10.

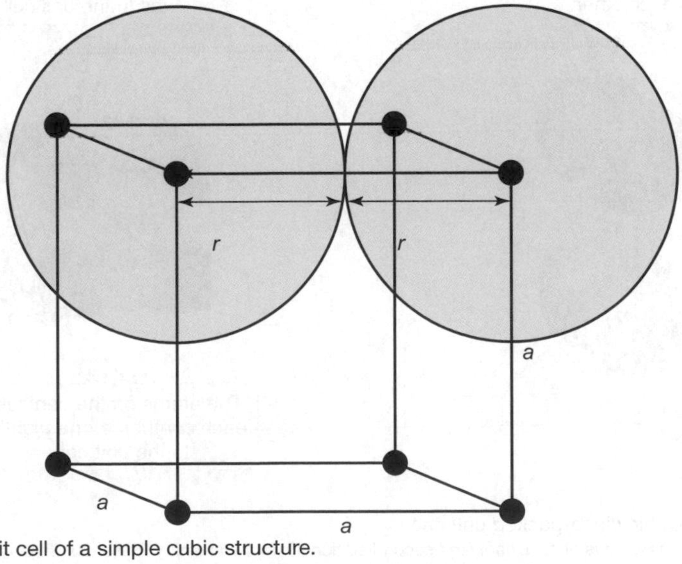

Figure 4.10. Unit cell of a simple cubic structure.

Only two atoms have been drawn in full along one of the sides of the unit cell. The radius of each atom is r and the length of the side of the unit cell is a.

The packing efficiency is given by:

$$\text{Packing efficiency} = \frac{\text{Volume occupied by atoms}}{\text{Total volume of space used}} \times 100\%.$$

We need to calculate the total volume of the unit cell that is occupied by atoms and also the volume of the unit cell:

$$\text{Packing efficiency} = \frac{V_{\text{Atoms}}}{V_{\text{Unit cell}}} \times 100\%$$

The volume of the unit cell is given by length × breadth × height of the shape—in this case a cube which has sides equal to a, thus the volume $= a^3$.

The volume of each atom is given by $4/3\,\pi r^3$. There are eight atoms involved in the unit cell but only one eighth of each atom is within a single unit cell. So the number of atoms in a unit cell $= 8 \times 1/8 = 1$ so the total volume occupied by these atoms $= 1 \times 4/3\,\pi r^3$.

Now we need to express the length of the unit cell (a) in terms of the radius of the atom (r). As the atoms are touching along the sides of the unit cell then we can see that the length of the side, $a = 2r$.

We can now substitute for a in the expression for the volume of the unit cell such that:
Volume of unit cell $= a^3 = (2r)^3 = 8r^3$.
Putting these expressions into the equation for packing efficiency gives:

The r^3 terms on the top and bottom of the expression cancel.

$$\text{Packing efficiency} = \frac{V_{\text{Atoms}}}{V_{\text{Unit cell}}} \times 100\% = \frac{\frac{4}{3}\pi r^3}{8r^3} \times 100\% = \frac{\pi}{3 \times 2} \times 100\% = 52.4\%$$

Worked example 4.1B

Calculate the packing efficiency in a cubic close-packed structure with a unit cell as shown in Figure 4.11.

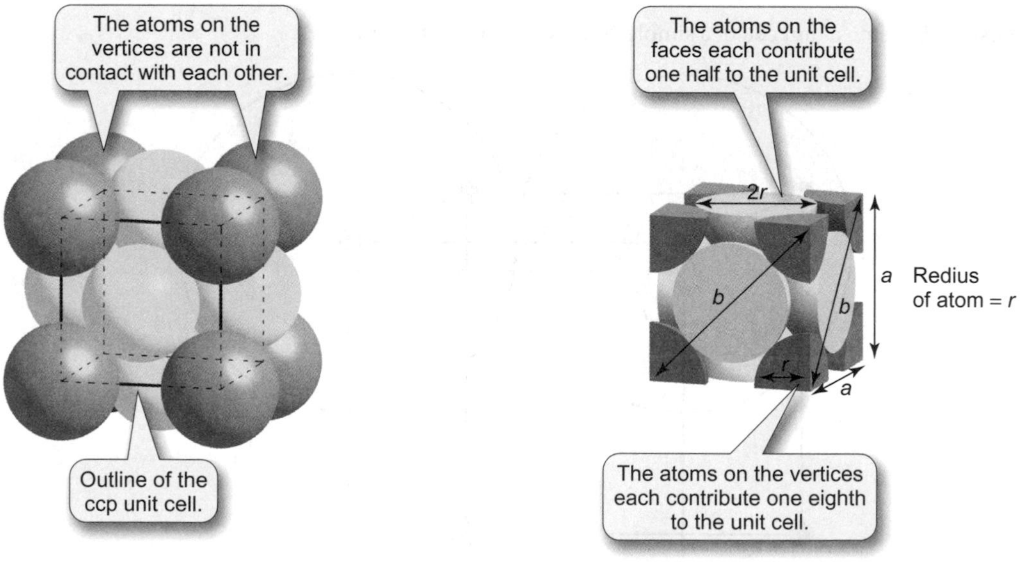
The atoms on the vertices are not in contact with each other.
Outline of the ccp unit cell.
The atoms on the faces each contribute one half to the unit cell.
2r
a
b
b
r
a
Redius of atom = r
The atoms on the vertices each contribute one eighth to the unit cell.

Figure 4.11 Cubic close-packed unit cell.
Reproduced from Burrows et al., *Chemistry*[3] second edition (Oxford University Press, 2013). © Andrew Burrows, John Holman, Andrew Parsons, Gwen Pilling, and Gareth Price 2013.

Solution

First of all sketch a unit cell of a cubic close-packed structure and label the unit cell length a and the radius of each atom, r, as in Figure 4.11. Label the length of the face diagonal b.

The packing efficiency is given by:

$$\text{Packing efficiency} = \frac{\text{Volume occupied by atoms}}{\text{Total volume of space used}} \times 100\%$$

so we need to calculate the total volume of the unit cell that is occupied by atoms and also the volume of the unit cell.

$$\text{Packing efficiency} = \frac{V_{\text{Atoms}}}{V_{\text{Unit cell}}} \times 100\%$$

The volume of the unit cell is the length × breadth × height of the shape—in this case a cube which has sides equal to a, thus the volume $= a^3$.

Next we must calculate the volume occupied by the atoms in the unit cell. In order to do this we must first calculate the number of whole atoms in the unit cell.

In a cubic close-packed (ccp) or face-centred cubic (fcc) structure, there are two different positions for atoms, as can be seen in Figure 4.11. There is one atom on each corner or vertex and because there are eight corners in a cube there are eight of this type of atom. However, only one eighth of each corner atom is in the unit cell. So the total number of corner atoms that are present in a single unit cell $= 8 \times 1/8 = 1$ atom.

The other position is in the centre of the faces of the cube, and there is one such atom on each face of the cube. A cube has six faces, and hence six of this type of atom. However, each of these atoms is shared by the adjoining unit cell such that only one half of the face atoms belong to a single unit cell.

Thus the total number of face atoms in a unit cell $= 6 \times ½ = 3$.

Therefore, the total number of whole atoms in one unit cell $= 1 + 3 = 4$.

The volume of an atom can be represented by the volume of a sphere $= 4/3\pi r^3$.

So the total volume occupied by the atoms in a ccp unit cell $= 4 \times 4/3\pi r^3$.

We now have the volume of the atoms and the volume of the unit cell, but the former is expressed in terms of r, the radius of each atom, and the latter is expressed in terms of a, the unit cell length. In order to get the packing efficiency of the cell we need to express a in terms of r.

In a ccp unit cell the atom on the centre of the face is in contact with the two diagonal corner atoms across the face. Thus the length of the face diagonal, $b = (r + 2r + r) = 4r$, as illustrated in Figure 4.12a.

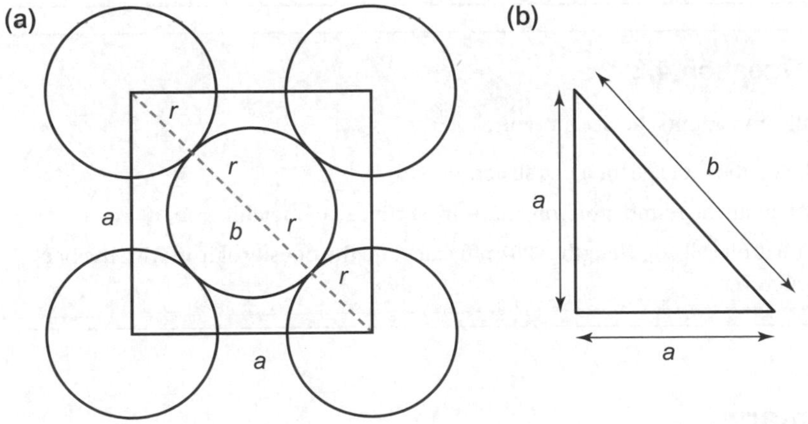

Figure 4.12 (a) A ccp unit cell. The length of the face diagonal, $b = (r + 2r + r) = 4r$. (b) A right-angled triangle which consists of two adjacent sides of the cube, length a, and the face diagonal, length b.

The face diagonal (b) is made up from two half atoms and one whole atom, each of radius r.

We can construct a right-angled triangle which consists of two adjacent sides of the cube, length a, and the face diagonal, length b (Figure 4.12b).

When we apply Pythagoras' theorem to this triangle we get the relationship: $a^2 + a^2 = b^2$.

Because $b = 4r$ we can substitute for b in this equation:

$$a^2 + a^2 = (4r)^2$$
$$2a^2 = (4r)^2$$

So $a^2 = \frac{(4r)^2}{2}$ therefore $a = \sqrt{\frac{(4r)^2}{2}} = \frac{4r}{\sqrt{2}}$

We can now calculate the volume of the unit cell in terms of r:

$$a^3 = \left(\frac{4r}{\sqrt{2}}\right)^3 = \frac{64r^3}{2\sqrt{2}} = \frac{32}{\sqrt{2}}r^3$$

Now we have a relationship between a and r we can substitute for a in the equation that allows us to calculate the packing efficiency:

$$\text{Packing efficiency} = \frac{V_{\text{Atoms}}}{V_{\text{Unit cell}}} \times 100\% = \frac{4 \times \frac{4}{3}\pi r^3}{a^3} \times 100\% = \frac{4 \times \frac{4}{3}\pi r^3}{\frac{32r^3}{\sqrt{2}}} \times 100\%$$

Simplifying the terms and cancelling r^3 on the top and bottom lines gives us:

$$\text{Packing efficiency} = \frac{\frac{16\pi}{3}}{\frac{32}{\sqrt{2}}} \times 100\% = \frac{\pi\sqrt{2}}{3 \times 2} \times 100\% = 74.0\%$$

Question 4.1

Calculate the packing efficiency in a body-centred cubic structure.

Question 4.2

Brass is an alloy of copper and zinc. A particular alloy of brass adopts a face-centred cubic structure in which the zinc atoms occupy the corners of the unit cell and the copper atoms occupy the centres of the faces. What is the mass in grams of a unit cell of this alloy of brass?

Question 4.3

Metallic iron adopts the bcc structure.

(a) Sketch the unit cell for a bcc structure.

(b) Calculate the number of iron atoms in a unit cell of metallic iron.

(c) If the unit cell edge length is 287 pm calculate the density of iron with the bcc structure.

Summary

From the above it should be clear that there are certain features of cubic structures which differ and are dependent upon whether we have a simple cubic, body-centred cubic, or cubic close-packed structure. These are summarized in Table 4.1.

Table 4.1 Characteristic features of different cubic structures.

Structure	Coordination number of atoms	Total number of atoms in unit cell	Packing efficiency	Feature of unit cell
Simple cubic	6	1	52.4%	Atoms are in contact along the side of unit cell
Body-centred cubic bcc	8	2	68.0%	Atoms are in contact along the body diagonal
Cubic close-packed ccp or fcc	12	4	74.0%	Atoms are in contact along the face diagonal

4.2 **Ionic lattices**

So far we have discussed solids in which all the spheres or atoms are the same. Such solids generally consist of metallic materials. However, the vast majority of solids are made up of atoms or groups of atoms of different types. Binary compounds are materials in which there are two types of atoms—generally of different sizes. Binary **ionic** compounds are materials made up of two types of ions with opposite charges and different sizes. Many ionic compounds can be thought of as consisting of one type of ion (usually the larger one) arranged in a close-packed array with the oppositely charged ions located in some of the interstitial holes left by the close-packed ions. Because there are two types of close-packed structures, hcp and ccp, and two types of interstitial holes, tetrahedral and octahedral, a variety of different ionic solid types can be derived from this type of packing.

In this section we will describe a few of the most common ionic structures that are based on close packing and look at the arrangements of the ions in these structures. This topic is much easier to understand if you use a model kit to build three-dimensional structures. Such structures allow you to envisage the coordination numbers of each ion more easily and also the repeating nature of ionic lattices.

The rock salt (sodium chloride) structure

The structure of sodium chloride and several other alkali metal halides is made up of a cubic close-packed (ccp) arrangement of chloride ions. These can be thought to occupy the corners of each unit cell and the centres of the faces (thus fcc) as for the metal atoms in section 4.1. The cations occupy all of the octahedral holes left by the chloride ions. The structure of a single unit cell of NaCl is shown in Figure 4.13. However, this unit is repeated throughout the whole of the

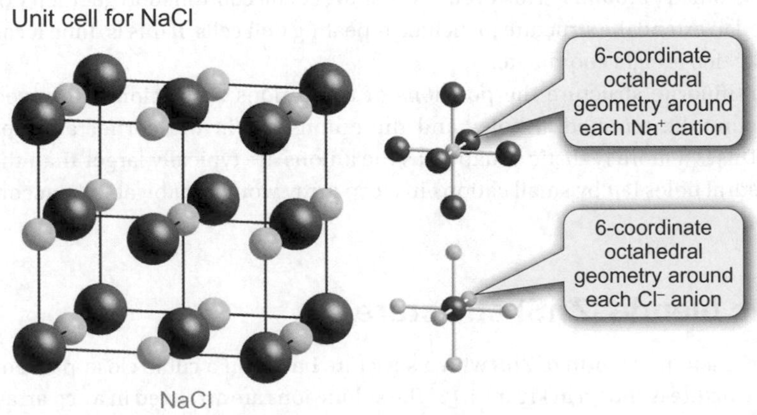

Figure 4.13 The rock salt (sodium chloride) structure.
Reproduced from Burrows et al., *Chemistry*[3] second edition (Oxford University Press, 2013). © Andrew Burrows, John Holman, Andrew Parsons, Gwen Pilling, and Gareth Price 2013.

crystalline solid. If you extend the unit cell in any direction you will see that the Cl^- ions and the Na^+ ions are actually in identical positions relative to each other. We can therefore also consider the structure as being a close-packed array of sodium ions with chloride ions in the octahedral holes and we can draw a unit cell with Na^+ ions on the corners and centres of the faces and Cl^- ions in the octahedral holes.

Each Na^+ ion is surrounded by six Cl^- ions and each Cl^- ion is surrounded by six Na^+ ions. So the coordination number of each ion is six.

The fluorite (CaF_2) and antifluorite structure

The unit cell of CaF_2 is shown in Figure 4.14.

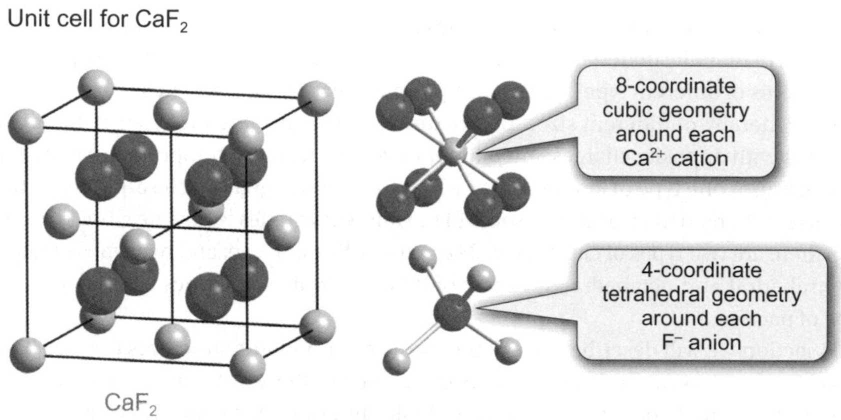

Figure 4.14 The fluorite structure.
Reproduced from Burrows et al., *Chemistry*[3] second edition (Oxford University Press, 2013). © Andrew Burrows, John Holman, Andrew Parsons, Gwen Pilling, and Gareth Price 2013.

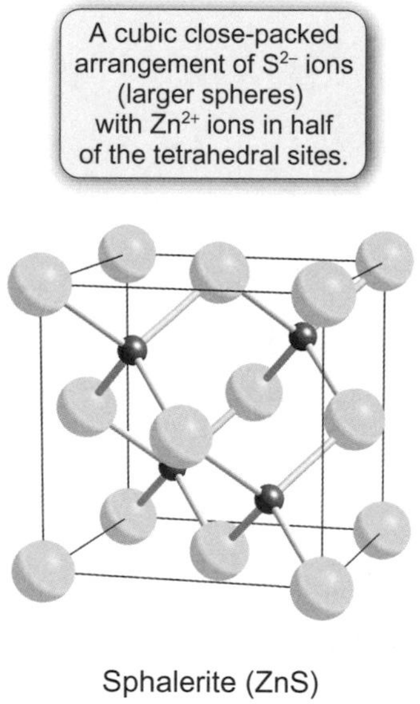

A cubic close-packed arrangement of S^{2-} ions (larger spheres) with Zn^{2+} ions in half of the tetrahedral sites.
Sphalerite (ZnS)

Figure 4.15 The structure of zinc blende (ZnS).
Adapted from Burrows et al., *Chemistry*[3] second edition (Oxford University Press, 2013). © Andrew Burrows, John Holman, Andrew Parsons, Gwen Pilling, and Gareth Price 2013.

Clearly the stoichiometry of this salt is 1:2 whereas NaCl has 1:1 stoichiometry. The fluorite structure can be thought of as a cubic close-packed array of Ca^{2+} ions with F^- ions in all the tetrahedral holes. There are two tetrahedral holes per Ca^{2+} ion and so the stoichiometry is 1:2. It is clear from the unit cell that each fluoride ion is four coordinate as each has four Ca^{2+} ions in a tetrahedral geometry around it. However, in order to see the coordination geometry of the Ca^{2+} ions we need to extend the structure to include repeating unit cells. If this is done it can be seen that each Ca^{2+} ion is eight coordinate.

In the antifluorite structure the positions of the cations and anions are reversed. The cations occupy the tetrahedral holes and the anions are in the corner and centre-face positions. This is a more realistic situation as the anions are typically larger than the cations and tetrahedral holes left by small cations in a ccp array would not be able to accommodate larger anions.

The zinc blende (ZnS) structure

Zinc blende is a mineral form of ZnS with a structure based on a cubic close-packed arrangement; this structure is shown in Figure 4.15. The sulfide ions are arranged in a ccp array with the zinc ions occupying half the tetrahedral holes.

Each zinc ion is tetrahedrally coordinated by four sulfide ions and each sulfide ion is tetrahedrally coordinated by four zinc ions. The stoichiometry is 1:1.

The wurtzite (zinc sulfide) structure

Wurtzite is another naturally occurring mineral form of ZnS but this material has a hexagonal close-packed structure, as shown in Figure 4.16. The sulfide ions are arranged as the ABAB layers in a hexagonal close-packed (hcp) array. The zinc ions are in the tetrahedral holes between the layers. The coordination number of both the Zn^{2+} and S^{2-} ions is four.

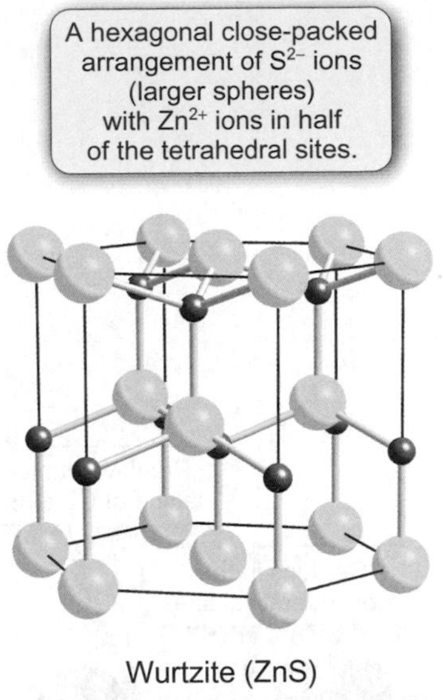

Figure 4.16 The structure of wurtzite (ZnS).
Adapted from Burrows et al., *Chemistry*³ second edition (Oxford University Press, 2013). © Andrew Burrows, John Holman, Andrew Parsons, Gwen Pilling, and Gareth Price 2013.

The cadmium chloride and cadmium iodide structures

$CdCl_2$ is based upon a cubic close-packed array of chloride ions with cadmium ions in half the octahedral holes, as shown in Figure 4.17a. The overall stoichiometry is 6:3 or 2:1 (anion:cation). The Cd^{2+} ions are hence six coordinate and the Cl^- ions are three coordinate.

CdI_2 is based upon a hcp array of iodide ions with Cd^{2+} ions in half the octahedral holes as shown in Figure 4.17b. Again each Cd^{2+} ion is six coordinate and each I^- ion is three coordinate.

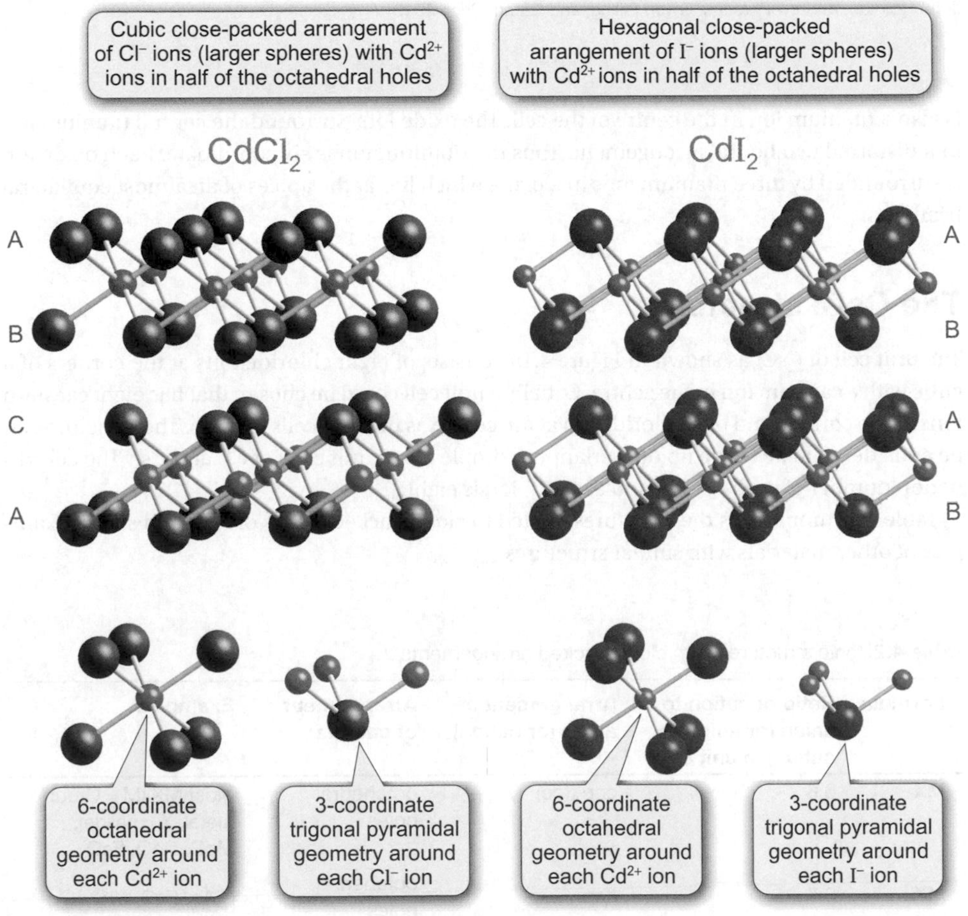

Figure 4.17 The structures of (a) $CdCl_2$ and (b) CdI_2.
Reproduced from Burrows et al., *Chemistry*³ second edition (Oxford University Press, 2013). © Andrew Burrows, John Holman, Andrew Parsons, Gwen Pilling, and Gareth Price 2013.

Zinc blende and wurzite are composed of the same atoms, Zn and S, arranged in different ways. The different arrangements of atoms in the solid state are different **polymorphs** of ZnS, and many other materials exhibit **polymorphic** behaviour. This is related to the phenomenon of allotropy, where some elements can exist in a number of solid forms, e.g. diamond and graphite, which are **allotropes** of carbon.

The rutile structure

The rutile structure is not a close-packed structure. It is named after one of the naturally occurring mineral forms of TiO_2. The unit cell is tetragonal—this means it is not cubic, but rectangular. As can be seen in Figure 4.18 the titanium ions occupy the corners of the unit cell and there

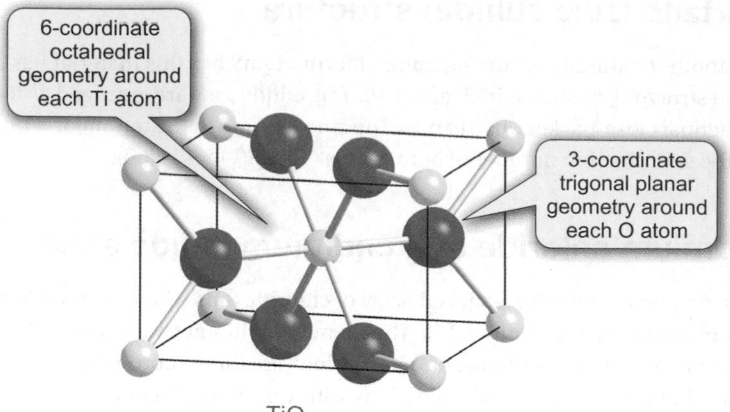

Figure 4.18 The rutile structure.
Reproduced from Burrows et al., *Chemistry*[3] second edition (Oxford University Press, 2013). © Andrew Burrows, John Holman, Andrew Parsons, Gwen Pilling, and Gareth Price 2013.

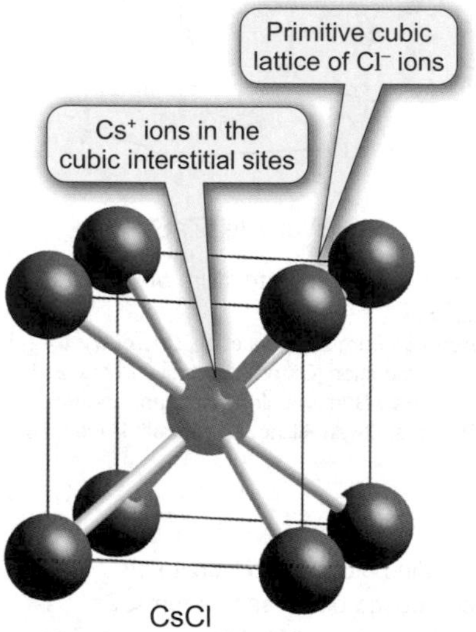

Figure 4.19 The caesium chloride structure.
Reproduced from Burrows et al., *Chemistry*[3] second edition (Oxford University Press, 2013). © Andrew Burrows, John Holman, Andrew Parsons, Gwen Pilling, and Gareth Price 2013.

is also a titanium ion in the centre of the cell. The oxide ions surround the central titanium ion in a distorted octahedral arrangement. Thus the titanium ion is six coordinate. Each oxide ion is surrounded by three titanium ions in a plane which lies at the apices of an almost equilateral triangle.

The CsCl structure

The unit cell of CsCl as shown in Figure 4.19 consists of eight chloride ions at the corners of a cube with a caesium ion in the centre. Equally a unit cell could be chosen that has eight caesium ions at the corners and one chloride ion at the centre as the unit cells overlap. The structure can be considered to be made up of overlapping simple cubic unit cells of Cl^- and Cs^+. The coordination number of each Cs^+ ion and each Cl^- ion is eight.

Table 4.2 summarizes the structures related to close-packed arrays of ions and shows examples of other materials with similar structures.

Table 4.2 Ionic structures with close-packed arrangements.

Formula	Ratio of cation to anion (or anion to cation) in unit cell	Arrangement of anion (or cation)	Arrangement of cations	Examples
MX	6:6	ccp (fcc)	octahedral holes	Most MX (M = Group 1 metal, X = halide), MgO, CaO, FeO
MX	6:6	hcp	octahedral holes	NiAs, FeS, NiS
MX	4:4	ccp	half tetrahedral holes	ZnS (zinc blende), CuCl
MX	4:4	hcp	half tetrahedral holes	ZnS (wurtzite), ZnO
MX_2	8:4	ccp	all tetrahedral holes	CaF_2, ZrO_2, $BaCl_2$
MX_2	6:3	ccp	half octahedral holes	$CdCl_2$, $MgCl_2$, TaS_2
MX_2	6:3	hcp	half octahedral holes	CdI_2, PbI_2, TiS_2

Worked example 4.2A

Draw the face-centred cubic structure of NaCl by placing Cl^- ions on the corners and centre of the faces of a cube and placing the Na^+ ions on the centres of the edges, with one Na^+ ion in the centre of the cell.

(a) What are the coordination numbers of the Na^+ and Cl^- ions in the NaCl structure?

(b) Calculate the number of formula units of NaCl in the unit cell by considering each position occupied by the Na^+ and Cl^- ions in turn and determining the fraction of each ion in a single unit cell.

Solution

The diagram to be drawn is shown in Figure 4.20.

(a) In a face-centred cubic structure such as this, both the Na^+ and the Cl^- ions are six coordinate (see Figure 4.13).

(b) To determine the numbers of each type of ion in a single unit cell we must consider each ion in turn and identify how many other unit cells each type of ion is shared by. Taking the Cl^- ions first:

There are eight corner Cl^- ions. Each is shared by a total of eight unit cells so one eighth of each corner ion is present in a single unit cell.

Total corner Cl^- ions $= 8 \times 1/8 = 1$ Cl^-.

There are six face Cl^- ions. Each is shared by two unit cells so one half of each face ion is present in a single unit cell.

Total face Cl^- ions $= 6 \times 1/2 = 3$ Cl^-.

Total Cl^- ions $= 1 + 3 = 4$.

Taking the Na^+ ions. There are 12 edges in a cube and one Na^+ ion at the centre of each edge. Each edge of a single unit cell is shared by a total of four unit cells. So there is a quarter of each Na^+ ion in a single unit cell.

Number of edge Na^+ ions $= 12 \times 1/4 = 3$ Na^+.

The remaining type of ion is the Na^+ at the centre. This is not shared by any other unit cells and so is wholly in a single unit cell.

Total Na^+ ions $= 3 + 1 = 4$.

So there are four Cl^- ions and four Na^+ ions in a unit cell making four NaCl formula units per cell.

► **Hint** We can summarize the procedure in a table and this provides a rigorous way of totalling the number of ions of each type in the unit cell:

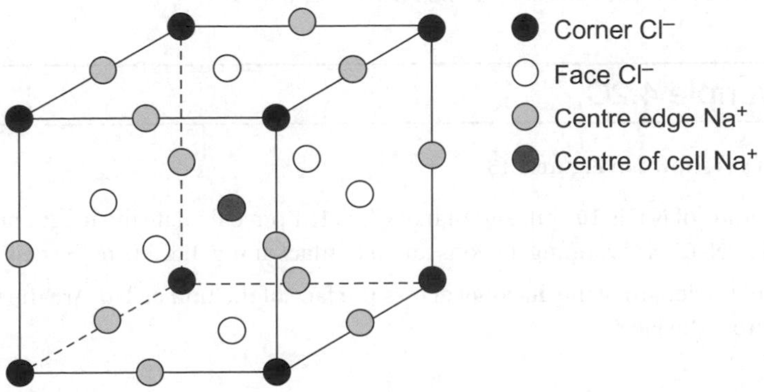

Figure 4.20 The structure of NaCl.

It is often difficult to visualize the number of adjoining unit cells in order to calculate the fraction of each type of ion in a single unit cell. Building three-dimensional models will really help you understand this.

Ion	Position	Number per unit cell	Fraction per unit cell	Total number in unit cell
Cl^-	corner	8	1/8	$8 \times 1/8 = 1$
Cl^-	face	6	1/2	$6 \times 1/2 = 3$
Total				$1 + 3 = 4$
Na^+	centre edge	12	1/4	$12 \times 1/4 = 3$
Na^+	centre cell	1	1	$1 \times 1 = 1$
Total				$3 + 1 = 4$

Worked example 4.2B

(a) Describe the positions of the Zn and S atoms in zinc blende.

(b) Give the coordination numbers of the Zn and S atoms in the structure of zinc blende (ZnS) as shown in Figure 4.15.

(c) Calculate the number of formula units in a unit cell of zinc blende.

Solution

(a) The larger sulfur atoms are in a face-centred cubic arrangement on the corners of the unit cell and the centres of the faces. The smaller zinc atoms occupy half the tetrahedral holes.

(b) It can be seen from the unit cell in Figure 4.15 that the coordination number of the Zn atoms is four. The coordination number of the S atoms is not as obvious. However, by taking one of the S atoms at the centre of a face, it can be seen that such a face atom is attached to two Zn atoms in the cell shown and another two Zn atoms in the adjoining cell. The coordination number of the S atoms is therefore also four.

(c) To find the total number of ZnS atoms in a unit cell we take the cation and anion in turn and determine how many other unit cells each is shared by. Set up a table as in Worked example 4.2A.

In this structure the S^{2-} ions are in the same positions as the Na^+ ions in NaCl (or the Cl^- ions) but in zinc blende the counter ions occupy half of the tetrahedral holes in the unit cell.

Ion	Position	Number per unit cell	Fraction per unit cell	Total number in unit cell
Zn^{2+}	centre cell	4	1	$4 \times 1 = 4$
Total				4
S^{2-}	corner	8	1/8	$8 \times 1/8 = 1$
S^{2-}	face	6	1/2	$6 \times 1/2 = 3$
Total				$1 + 3 = 4$

You can check that the ratio is correct as the charge on Zn = +2 and the charge on S = −2 so the charge is balanced.

Zn = 4 and S = 4 so the total number of ZnS formula units = 4.

Worked example 4.2C

A unit cell of NaCl is shown in Figure 4.13.

(a) If the ionic radius of Na^+ is 102 pm and that of Cl^- is 181 pm calculate the length of the side of a unit cell in NaCl, a, assuming the ions are in contact along the side of the cell.

(b) Next work out the length of the diagonal across the face of the unit cell, d. Are the Cl^- ions in contact across the face?

Solution

(a) The unit cell length in NaCl is made up of one ion of Na^+ and two half ions of Cl^-. Thus if the ions are in contact:

the length of the side $= 2 \times r(Cl^-) + 2 \times r(Na^+)$

$= 2 \times 181\ pm + 2 \times 102\ pm$

$= 566\ pm$

Applying Pythagoras' theorem, the length of the face diagonal, *d*, can be determined by using the right-angled triangle that has two cell edges (*a*) and the diagonal (*d*) as the hypotenuse.

So $a^2 + a^2 = d^2$

$a = 566\ pm$

$(566)^2 + (566)^2 = 640712\ pm^2 = d^2$

$d = 800\ pm.$

(b) To determine if the ions are in contact calculate the distance occupied by four Cl^- radii $= 4 \times 181\ pm = 724\ pm$.

So the ions are not in contact.

If the unit cell was drawn with Na^+ ions on the corners and Cl^- ions on the centre of the edges the unit cell length would still be the same, *i.e.* the radius of two sodium ions and two chloride ions.

We would not expect Cl^- ions to be in contact as they are negatively charged.

Worked example 4.2D

Calculate the density of NaCl if there are four NaCl formula units per unit cell and the unit cell length is 566 pm.

Solution

The density of a solid is determined from the mass/volume:

$$\text{Density} = \frac{\text{Mass}}{\text{Volume}}$$

$$\rho = \frac{m}{V}$$

The symbol for density is ρ (rho). Sometimes the symbol *d* is used.

There are four formula units of NaCl per unit cell. The mass of one formula unit can be determined from the atomic masses:

$Na = 22.99\ g\ mol^{-1}$, $Cl = 35.45\ g\ mol^{-1}$.

The mass of one mole of formula unit is therefore $58.44\ g\ mol^{-1}/N_A$:

$= 58.44\ g\ mol^{-1}/6.022 \times 10^{23}\ mol^{-1}$.

Therefore the mass of four moles of formula units $= 4 \times 58.44\ g\ mol^{-1}/6.022 \times 10^{23}\ mol^{-1} = 38.82 \times 10^{-23}\ g$.

The volume of a unit cell of NaCl $= a^3$ where $a = 566$ pm = unit cell length.

So $V = (566\ pm)^3 = 1.813 \times 10^8\ pm^3$.

Density = mass/volume

$= 38.82 \times 10^{-23}\ g/1.813 \times 10^8\ pm^3$

$= 21.41 \times 10^{-31}\ g\ pm^{-3}$

$= 21.41 \times 10^{-31} \times 10^{30}\ g\ cm^{-3}$

$= 2.141\ g\ cm^{-3}$

To convert from pm to cm: 1 pm = 10^{-10} cm.

Question 4.4

(a) A unit cell of calcium fluoride is shown in Figure 4.21. It has the fluorite structure. Describe the structure in terms of the arrangement of the cations and anions, the occupancy of the holes, and the coordination numbers of the ions.

(b) Determine the number of CaF_2 formula units in a unit cell of calcium fluoride by considering the occupancy of each ion in the cell.

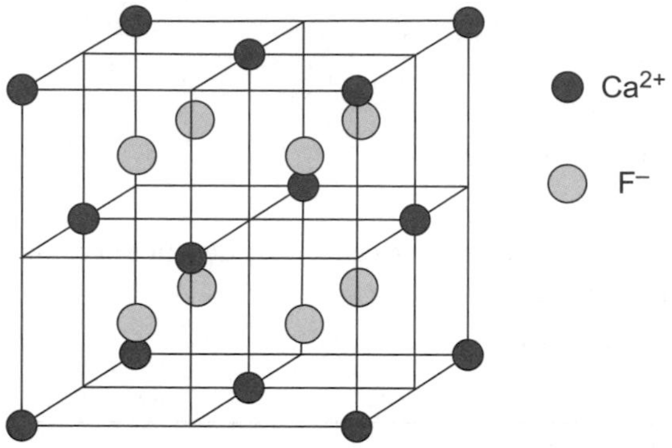

Figure 4.21 A unit cell of calcium fluoride.

Question 4.5

Crystalline cadmium oxide (CdO) has a density of 8.2 g cm^{-3} and adopts the sodium chloride structure. If the edge of the unit cell is 470 pm, how many Cd^{2+} and O^{2-} ions are there in the unit cell? (Relative formula mass of CdO = 128).

4.3 The radius ratio rule

The radius ratio rule predicts the structure that an ionic solid will adopt, based upon the ratio of the cation to anion size. The basic idea is that the anions in a crystalline lattice are arranged regularly and the cations (which are generally smaller) fit into interstitial 'holes' (either tetrahedral or octahedral). The most stable lattice will be formed when the cation can maximize its interactions with the anion or have the largest number of nearest neighbours.

If the cation can touch all its neighbours then the electrostatic interactions will be maximized as in Figure 4.22a. If the cation size is any smaller for the same size anion then the hole will be too big, the cation–anion interactions will be reduced and the anions will touch as in Figure 4.22b. This is not favoured electrostatically and so the anions will rearrange to a different structure type with a smaller interstitial hole and lower coordination number. If the cation size increases then the size of the interstitial hole made by the anion must be large enough to both accommodate the cation and maximize the interactions (Figure 4.22c). If the cation gets too large then the anions must rearrange themselves to a packing arrangement with a larger interstitial hole size and higher coordination number.

The radius ratio is the ratio of the smaller ion (usually the cation) to the larger ion (usually the anion), and hence always results in a value less than 1.

$$\text{Radius ratio} = \frac{\text{Radius of cation}}{\text{Radius of anion}} = \frac{r_+}{r_-}$$

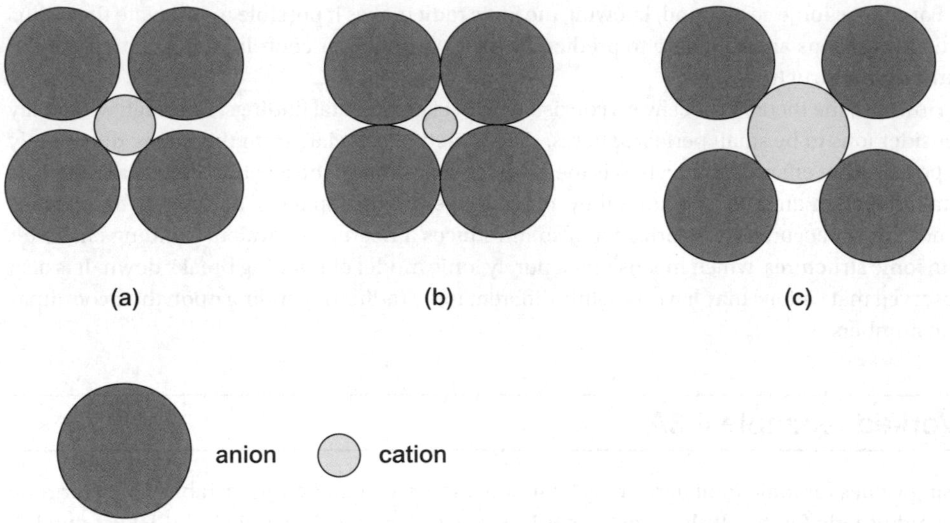

Figure 4.22 The radius ratio rule.

The size of the interstitial hole depends upon the radius of the anion and the coordination type. We have already seen that tetrahedral holes are smaller than octahedral holes and so can only accommodate small cations. Trigonal holes (with a coordination number of three), although not very common, are smaller than tetrahedral holes. Cubic holes (with a coordination number of eight) are larger as can be envisaged. The limiting ratio of the cation size to the anion size can be calculated using geometry for each of these coordination types. This is called the limiting radius ratio for each coordination geometry. These are shown in Table 4.3.

Table 4.3 Limiting radius ratios for different coordination geometries.

Coordination number	Geometry around cation	Limiting radius ratio = r_{cation}/r_{anion}
2	linear	0–0.155
3	trigonal	0.155–0.225
4	tetrahedral	0.225–0.414
6	octahedral	0.414–0.732
8	cubic	0.732–1.0
12	cuboctahedral	1.0

Table 4.4 Crystallographically determined radii (pm) of selected ions.

Ion	Radius	Ion	Radius	Ion	Radius	Ion	Radius	Ion	Radius
Li^+	90	Be^{2+}	59	Fe^{2+}	75/92†	O^{2-}	126	F^-	119
Na^+	116	Mg^{2+}	86	Co^{2+}	79/89†	S^{2-}	170	Cl^-	167
K^+	152	Ca^{2+}	114	Ni^{2+}	69/63*	Se^{2-}	184	Br^-	182
Rb^+	166	Sr^{2+}	132	Cu^{2+}	87			I^-	206
Cs^+	181	Ba^{2+}	149	Zn^{2+}	74/88‡			OH^-	123
				Ag^+	129			N^{3-}	132

†low spin/high spin
‡tetrahedral/octahedral
*tetrahedral/square planar
Source: Shannon, R. D. (1976) 'Revised Effective Ionic Radii and Systematic Studies of Interatomic Distances in Halides and Chalcogenides', *Acta Cryst.*, A32, 751–767.

There are many different ways of determining ionic radii as these values are not precisely defined. Ions can be squashed and distorted such that their size and shape depend upon the environment. Several sets of ionic radii have been determined; although each set is internally consistent, the values from different sets of radii should not be mixed. A description of the different ways of determining ionic radii can be found in Smart and Moore (2012). A comprehensive set of ionic radii has been compiled by Shannon and Prewitt (using electron density maps obtained from X-ray crystallography experiments on many ionic crystal structures) (Shannon, 1976). Values for some common ions are given in Table 4.4. This set of radii tend to make the anions smaller and cations bigger.

For a given ionic compound, knowing the ionic radii makes it possible to calculate the radius ratio for the ions and therefore to predict the likely structure or coordination geometry of the cation in the structure.

However, the theory is not always consistent with experimental findings. Although we usually consider ions to be small perfect spheres, they sometimes deviate from this shape due mainly to polarization effects. Polarization is the result of distortion of the electron clouds around certain ions (often anions), such that they are no longer regular spheres, and hence their radius is not known accurately. Polarization also introduces a degree of covalent bonding character into ionic structures, which means that a purely ionic model of bonding breaks down. It is also observed that cations may have a slightly different ionic radius depending upon their coordination numbers.

Worked example 4.3A

Using values for ionic radii derived by Shannon and Prewitt and given in Table 4.4, determine the radius ratio for the alkali metal halides NaF, NaCl, NaBr, and NaI and LiF, LiCl, LiBr, and LiI. Compare the predicted geometry with the octahedral geometry found by X-ray crystallography.

Solution

For the sodium halides the radius ratio is given by $= \dfrac{116\ \text{pm}}{r(\text{X}^-)}$

$\text{NaF} = \dfrac{116\ \text{pm}}{119\ \text{pm}} = 0.975$ Therefore, predict cubic, actual geometry = octahedral.

$\text{NaCl} = \dfrac{116\ \text{pm}}{167\ \text{pm}} = 0.695$ Therefore, predict octahedral, actual geometry = octahedral.

$\text{NaBr} = \dfrac{116\ \text{pm}}{182\ \text{pm}} = 0.637$ Therefore, predict octahedral, actual geometry = octahedral.

$\text{NaI} = \dfrac{116\ \text{pm}}{206\ \text{pm}} = 0.563$ Therefore predict octahedral, actual geometry = octahedral.

For the lithium halides the radius ratio is given by $= \dfrac{90\ \text{pm}}{r(\text{X}^-)}$.

$\text{LiF} = \dfrac{90\ \text{pm}}{119\ \text{pm}} = 0.756$ Therefore predict cubic, actual geometry = octahedral.

$\text{LiCl} = \dfrac{90\ \text{pm}}{167\ \text{pm}} = 0.539$ Therefore predict octahedral, actual geometry = octahedral.

$\text{LiBr} = \dfrac{90\ \text{pm}}{182\ \text{pm}} = 0.495$ Therefore predict octahedral, actual geometry = octahedral.

$\text{LiI} = \dfrac{90\ \text{pm}}{206\ \text{pm}} = 0.437$ Therefore predict octahedral, actual geometry = octahedral.

So the radius ratio rule using Shannon radii works well to predict all of these alkali halide geometries apart from NaF and LiF. This highlights the fact that the radius ratio rule is a very simple predictive model that doesn't account for all the factors that determine packing in ionic structures, other than the packing efficiency related to different sphere sizes. While it performs well for a large number of binary ionic compounds there are many exceptions.

Question 4.6

State whether the following ionic structures obey the radius ratio rule:

(a) Caesium bromide (cubic geometry).

(b) Zinc oxide (tetrahedral geometry).

(c) Copper oxide (tetrahedral geometry).

(d) Silver chloride (octahedral geometry).

(e) Magnesium fluoride (rutile structure with octahedrally coordinated Mg^{2+} ions).

4.4 Lattice enthalpy and the Born–Haber cycle—an experimental method for determining lattice enthalpy

The electrostatic attraction between cations and anions provides the force to hold ionic solids together in their crystal lattices. The strength of these forces is measured by the **lattice enthalpy** which we consider to be the enthalpy change that accompanies the formation of one mole of the crystalline lattice from its constituent gaseous ions. It is the enthalpy change for the process:

$$x\mathrm{M}^{m+}(\mathrm{g}) + y\mathrm{A}^{n-}(\mathrm{g}) \rightarrow \mathrm{M}_x\mathrm{A}_y(\mathrm{s})$$

In thermodynamic notation, the lattice enthalpy is given the shorthand $\Delta_{\mathrm{lat}}H^{\ominus}(M_xA_y)$.

When defined in this way, the lattice enthalpy is sometimes called the lattice enthalpy for formation, and always has an exothermic (i.e. negative) value. Other text books and data tables define the lattice enthalpy as the lattice enthalpy of dissociation, which is the reverse of the above process. The lattice enthalpy of dissociation is therefore the energy **required** to separate one mole of the solid into its constituent ions in the gas phase and as such is always endothermic (i.e. positive). The numerical value in both cases is identical; however, the sign changes depending on the definition used.

It is impossible to carry out an experiment to determine the lattice enthalpy directly from experiment and so the lattice enthalpy for a solid can only be determined indirectly from direct thermochemical measurements and through the use of an appropriate energy cycle.

We can consider the reaction between a metal and a non-metal to form an ionic solid (e.g. the reaction of sodium with chlorine or of magnesium with oxygen) as occurring in a number of discrete steps. These steps are:

i. Conversion of the metal into gaseous atoms—**the enthalpy of atomization**. This is defined as the enthalpy change associated with the formation of one mole of gaseous atoms from the element in its standard state. The enthalpy of atomization has the symbol $\Delta_{\mathrm{at}}H^{\ominus}$.

$$\mathrm{M(s)} \rightarrow \mathrm{M(g)}$$

ii. Conversion of the non-metal into one mole of gaseous atoms (again, this is the **enthalpy of atomization**). For the specific example of a diatomic molecule, this is equivalent to half of the **bond dissociation enthalpy**.

$$\tfrac{1}{2}\mathrm{X}_2(\mathrm{g}) \rightarrow \mathrm{X(g)}$$

Remember, the enthalpy of atomization is defined with respect to the formation of **one mole** of gaseous atoms and so dissociating half a mole of diatomic molecules results in one mole of monatomic atoms.

iii. Ionization of the metal to form cations; this may simply be the first ionization enthalpy if the metal is a +1 cation or a combination of ionization enthalpies (e.g. first and second ionization enthalpies to form M^{2+}).

$$\mathrm{M(g)} \rightarrow \mathrm{M}^{n+}(\mathrm{g}) + ne^-$$

iv. Addition of electrons to the non-metal to form anions. This is known as electron affinity or electron gain enthalpy and again, may simply be the first electron affinity for an X^- anion or a combination of electron affinities (e.g. first plus second electron affinities for an X^{2-} anion).

$$\mathrm{X(g)} + ne^- \rightarrow \mathrm{X}^{n-}(\mathrm{g})$$

The electron affinity, $\Delta_{\mathrm{EA}}H$, is the enthalpy change associated with the attachment of an electron to a gaseous atom of an element.

Confusingly a number of different names for various energy terms are used rather loosely and interchangeably in the chemical literature. Thus you may see the use of ionization enthalpies, ionization energies, and ionization potentials. Likewise electron affinities and electron gain enthalpies are used. For a metal like sodium the terms atomization enthalpy and sublimation enthalpy refer to the same chemical process, i.e. the conversion of sodium metal into gaseous sodium atoms. If you are unfamiliar with a particular term then try to find out the exact chemical process that the term corresponds to before using the value in a Born–Haber calculation.

In a Born–Haber cycle all of the terms used should be enthalpies (Δ*H* values). However, occasionally internal energy changes (Δ*U*) are used, as the values are more readily available. In general, the differences between Δ*H* and Δ*U* values tend to be small, so this discrepancy should not matter except for very accurate work.

Using the arrows to determine if we add or subtract a quantity is independent of whether a quantity is exothermic or endothermic. Sensible use of brackets can help to ensure calculations account for negative exothermic terms correctly.

v. Combination of the gaseous cations and anions to form the ionic solid. This is the **lattice enthalpy** and is defined here as an exothermic term for the combination of gaseous ions. This quantity is impossible to measure experimentally and hence is calculated using Born–Haber cycles or other theoretical methods.

$$M^{n+}(g) + X^{n-}(g) \rightarrow MX(s)$$

Great care must be taken in making sure that the mathematical signs used for the enthalpy terms in a Born–Haber cycle are correct. All endothermic terms should have a '+' sign and all exothermic terms should have a '–' sign. It is often useful to carry out a 'sanity check' to make sure that the signs you have used appear to be correct. For example atomization enthalpies and ionization enthalpies are always endothermic as these processes require energy.

An example of a generalized Born–Haber cycle for a salt, MX is given in Figure 4.23.

To use this cycle to calculate the lattice enthalpy for the salt MX, we must make use of the other known quantities in the cycle. Hess's law tells us that the enthalpy change for a process is independent of the route taken, and hence we can use the other quantities in this cycle (which can be obtained experimentally) to determine the lattice enthalpy. Thus the enthalpy change of formation of the salt, MX, represented by $\Delta_f H^{\ominus}$, is equivalent to the sum of the enthalpy changes obtained by travelling in the opposite direction, represented by the dashed arrow, from M(s) and $1/2X_2$(g) to MX(s) as given by the equation: $\Delta_f H^{\ominus}(MX) = \Delta_{at}H^{\ominus}(M) + IE_1(M) + \Delta_{at}H^{\ominus}(X) + \Delta_{EA}H^{\ominus}(X) + \Delta_{lat}H^{\ominus}(MX)$.

The equation can be rearranged to get an expression for the lattice enthalpy (exothermic, lattice formation definition) as:

$$\Delta_{lat}H^{\ominus}(MX) = -\Delta_{EA}H^{\ominus}(X) - \Delta_{at}H^{\ominus}(X) - IE_1(M) - \Delta_{at}H^{\ominus}(M) + \Delta_f H^{\ominus}(MX)$$

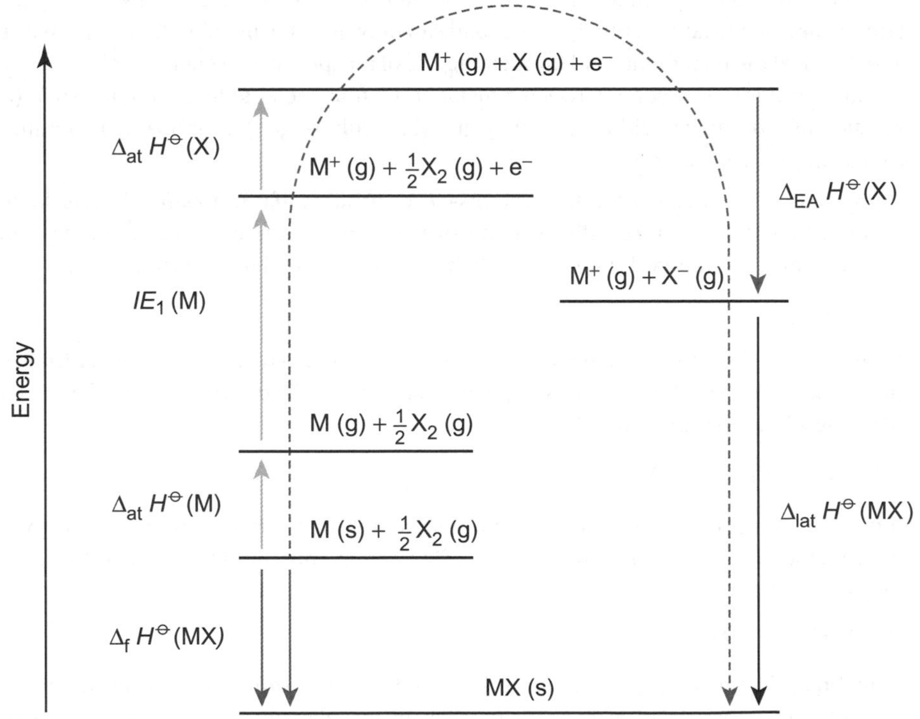

Figure 4.23 Born–Haber cycle for an ionic salt MX.

Worked example 4.4A

Use the data in the table below to calculate the lattice enthalpy for CsBr, by constructing a suitable Born–Haber cycle.

Quantity	Value/kJ mol^{-1}
Sublimation enthalpy for Cs	+78
1st ionization enthalpy for Cs	+375.5
Bond dissociation enthalpy for Br_2	+193.9
1st electron affinity for Br	–324
Enthalpy of formation of CsBr	–405.9

Solution

The first thing to do when answering a question of this sort is to determine the physical processes for which we have data. Most of the time, this is simple. However, occasionally non-standard terms are used which can complicate the process. The table below contains the physical processes associated with each of the quantities given in the question:

Quantity	Thermodynamic notation	Physical process	Value/kJ mol^{-1}
Sublimation enthalpy for Cs	$\Delta_{sub}H^{\ominus}(Cs)$	$Cs(s) \rightarrow Cs(g)$	+78
1st ionization enthalpy for Cs	$IE_1(Cs)$	$Cs(g) \rightarrow Cs^+(g) + e^-$	+375.5
Bond dissociation enthalpy for Br_2	$D_{(Br-Br)}$	$Br_2(g) \rightarrow 2Br(g)$	+193.9
1st electron affinity for Br	$\Delta_{EA}H^{\ominus}(Br)$	$Br(g) + e^- \rightarrow Br^-(g)$	–324
Enthalpy of formation of CsBr	$\Delta_f H^{\ominus}(CsBr)$	$Cs(s) + \frac{1}{2}Br_2(g) \rightarrow CsBr(s)$	–405.9

We can see that the sublimation enthalpy for caesium is exactly equivalent to the enthalpy of atomization for caesium. We can also see that the bond dissociation enthalpy for Br_2 is exactly double the enthalpy of atomization for bromine, and hence we will need to account for this in our calculations.

We can now use this information to construct the Born–Haber cycle shown in Figure 4.24. The energy axis in Figure 4.24 has been omitted for clarity.

As with the general example given in the introduction to this section, the quantity we wish to determine, the lattice enthalpy for CsBr, is given by the solid black arrow. Again, the enthalpy cycle allows us to equate the enthalpy change of formation of CsBr to the sum of the enthalpy changes obtained by travelling in the clockwise direction from Cs(s) and $1/2Br_2(g)$ to solid CsBr, i.e. by summing the enthalpy changes along the clockwise dashed line. By doing this we get the equation:

$$\Delta_f H^{\ominus}(CsBr) = \Delta_{at}H^{\ominus}(Cs) + IE_1(Cs) + \Delta_{at}H^{\ominus}(Br) + \Delta_{EA}H^{\ominus}(Br) + \Delta_{lat}H^{\ominus}(CsBr)$$

The equation can be rearranged to get an expression for the lattice enthalpy as follows:

$$\Delta_{lat}H^{\ominus}(CsBr) = -\Delta_{EA}H^{\ominus}(Br) - \Delta_{at}H^{\ominus}(Br) - IE_1(Cs) - \Delta_{at}H^{\ominus}(Cs) + \Delta_f H^{\ominus}(CsBr)$$

$$\Delta_{lat}H^{\ominus}(CsBr) = -(-324) - \left(\frac{1}{2}\times 193.9\right) - (375.5) - (78) + (-405.9)$$

$$\Delta_{lat}H^{\ominus}(CsBr) = +324 - 96.95 - 375.5 - 78 - 405.9$$

$$\Delta_{lat}H^{\ominus}(CsBr) = -632 \text{ kJ mol}^{-1}$$

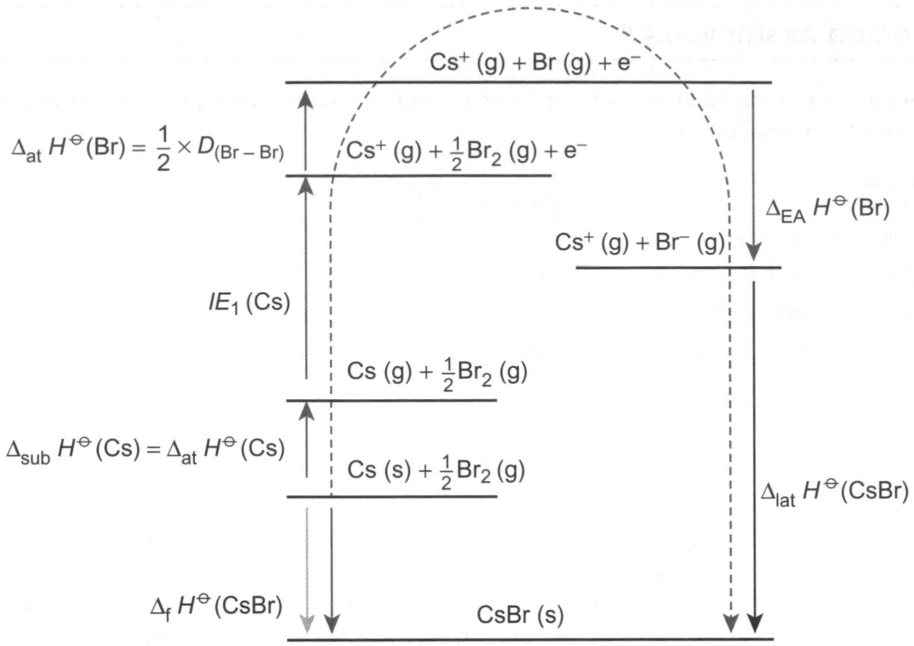

Figure 4.24 Born–Haber cycle for CsBr.

Worked example 4.4B

Given the following information for strontium, oxygen, and strontium oxide, calculate the second electron affinity for oxygen:

$$O^{-}(g)+e^{-} \rightarrow O^{2-}(g)$$

Quantity	Value/kJ mol^{-1}
Sublimation enthalpy for Sr	+164
1st ionization enthalpy for Sr	+549.5
2nd ionization enthalpy for Sr	+1064.2
Bond dissociation enthalpy for O_2	+499
1st electron affinity for O	–141
Lattice enthalpy for SrO	–3223
Enthalpy of formation of SrO	–592

Solution

As with Worked example 4.4A, the first thing we need to do when answering a question of this sort is to determine the physical processes that match the data we have been given. Most of the time, this is simple however, occasionally non-standard terms are used.

The table below contains the physical processes associated with each of the quantities given.

Quantity	Physical process	Value/kJ mol^{-1}
Sublimation enthalpy for Sr	$Sr(s) \rightarrow Sr(g)$	+164
1st ionization enthalpy for Sr	$Sr(g) \rightarrow Sr^{+}(g) + e^{-}$	+548
2nd ionization enthalpy for Sr	$Sr^{+}(g) \rightarrow Sr^{2+}(g) + e^{-}$	+1060
Bond dissociation enthalpy for O_2	$O_2(g) \rightarrow 2O(g)$	+496
1st electron affinity for O	$O(g) + e^{-} \rightarrow O^{-}(g)$	−142
Lattice enthalpy for SrO	$Sr^{2+}(g) + O^{2-}(g) \rightarrow SrO(s)$	−3310
Enthalpy of formation of SrO	$Sr(s) + \frac{1}{2}O_2(g) \rightarrow SrO(g)$	−590

We can see that the sublimation enthalpy for strontium is equivalent to the atomization enthalpy for strontium, and that the enthalpy of atomization for oxygen is equivalent to half of the bond dissociation enthalpy for oxygen.

We can now use this information to construct a Born–Haber cycle. There are more steps in this cycle than the one seen in Worked example 4.4A, because we now have two ionization events to form the Sr^{2+} cation, and two electron gain events to form the O^{2-} anion. It is the second of these that we have been asked to calculate. We can predict that this will be an endothermic term, due to repulsion between the O^{-} anion and e^{-}. Note that we do not need to know this in advance in order to calculate the correct value. A Born–Haber cycle for SrO is given in Figure 4.25. The energy axis has been omitted for clarity.

The quantity we have been asked to determine is highlighted by the solid black arrow. Again our enthalpy cycle allows us to equate the enthalpy of formation of SrO directly from the elements with the sum of the enthalpy changes following the dashed line in a clockwise direction from the solid Sr and gaseous molecules of O_2 though the gaseous ions to solid SrO. The expression for this is given by:

$$\Delta_f H^{\ominus}(\text{SrO}) = \Delta_{at}H^{\ominus}(\text{Sr}) + IE_1(\text{Sr}) + IE_2(\text{Sr}) + \Delta_{at}H^{\ominus}(\text{O}) + \Delta_{EA1}H^{\ominus}(\text{O}) + \Delta_{EA2}H^{\ominus}(\text{O}) + \Delta_{lat}H^{\ominus}(\text{SrO}).$$

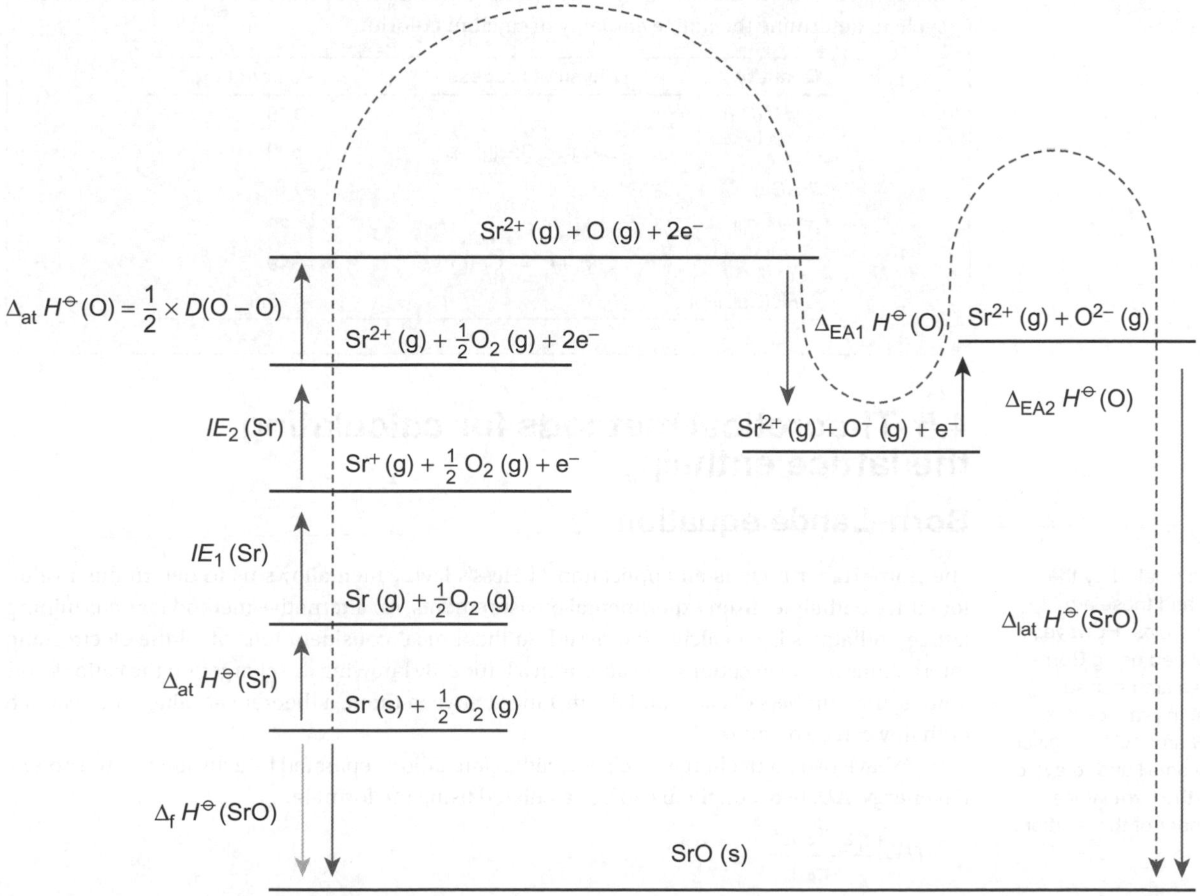

Figure 4.25 Born–Haber cycle for SrO.

This equation can be rearranged to obtain an expression for the second electron affinity of oxygen as follows:

$$\Delta_{EA2}H^{\ominus}(O) = -\Delta_{EA1}H^{\ominus}(O) - \Delta_{at}H^{\ominus}(O) - IE_2(Sr) - IE_1(Sr)$$
$$-\Delta_{at}H^{\ominus}(Sr) + \Delta_f H^{\ominus}(SrO) - \Delta_{lat}H^{\ominus}(SrO)$$

$$\Delta_{EA2}H^{\ominus}(O) = -(-142) - \left(\frac{1}{2}\times 496\right) - (1060) - (548) - (164) + (-590) - (-3310)$$

$$\Delta_{EA2}H^{\ominus}(O) = +142 - 248 - 1060 - 548 - 164 - 590 + 3310$$

$$\Delta_{EA2}H^{\ominus}(O) = +842 \text{ kJ mol}^{-1}$$

Question 4.7

NaCl, NaBr, and MgO all have the same crystal structure. Explain the following observations:

(a) The lattice enthalpy of NaBr is a little less than that of NaCl.

(b) The lattice enthalpy of MgO is about five times greater than that of NaCl.
(NaCl = 787 kJ mol^{-1}; NaBr = 747 kJ mol^{-1}; MgO = 3791 kJ mol^{-1}).

(c) The enthalpies of formation of NaCl and NaBr show a smaller variation from the value for MgO than do the lattice enthalpies (NaCl = 411 kJ mol^{-1}; NaBr = 360 kJ mol^{-1}; MgO = 602 kJ mol^{-1})

Question 4.8

Complete the table below, and use the values it contains to create a suitable Born–Haber cycle to determine the lattice enthalpy of calcium chloride.

Quantity	Physical process	Value/kJ mol^{-1}
$\Delta_{at}H^{\ominus}(Ca)$		+178
	$Ca(g) \rightarrow Ca^{+}(g) + e^{-}$	+590
$IE_2(Ca)$		+1146
$\Delta_{at}H^{\ominus}(Cl)$		+122
	$Cl(g) + e^{-} \rightarrow Cl^{-}(g)$	−349
$\Delta_f H^{\ominus}(CaCl_2)$		−795.8

4.5 Theoretical methods for calculating the lattice enthalpy

Born–Landé equation

Values of ΔU are calculated by the Born-Landé equation, and these are referred to as lattice *energies*. However, values of ΔH are calculated using Born-Haber cycles, and these are referred to as lattice *enthalpies*. In practice, the difference between ΔH and ΔU for a given process is usually very small and to avoid confusion, we will use the term lattice enthalpy for the remainder of this section.

The Born–Haber cycle is an application of Hess's law, which allows us to determine a value for lattice enthalpies from experimental measurements. An alternative method for determining lattice enthalpies is by calculation based on theoretical considerations of all the electrostatic interactions between cations and anions in a lattice. By knowing the charges on the cations and anions, the numbers of each, and the distance between them, a theoretical value for the lattice enthalpy can be obtained.

If we have two point charges such as a cation and anion separated by a distance, *r*, the potential energy, ΔU, between them can be calculated using the formula:

$$\Delta U = -\frac{|z_+||z_-|e^2}{4\pi\varepsilon_0 r}$$

where $|z_+|$ and $|z_-|$ are the magnitudes of the cation and anion charges respectively, e is the charge on an electron, ε_0 is the permittivity of vacuum, and r is the distance between the ions. The calculated energy term is negative as energy is released when the gaseous ions are brought together from an infinite distance.

This expression considers only two oppositely charged ions but in a crystal lattice there are huge numbers of attractive and repulsive forces between oppositely and similarly charged ions respectively. To compensate for this, the above energy term is adjusted by a constant called the Madelung constant, A, which is specific for each structural type (e.g. NaCl, CsCl, ZnS, etc.). The Madelung constant takes into account the specific crystal lattice and the relative electrostatic interactions between each of the cations and anions in their different positions.

Accounting for these interactions by A and converting to obtain the molar lattice energy by multiplication with the Avogadro constant, the equation becomes:

$$\Delta U = -N_A \times A \times \frac{|z_+||z_-|e^2}{4\pi\varepsilon_0 r}$$

In some text books, the Avogadro number is represented by the letter L. It does not really matter which symbol you adopt in your own work; just remember to use consistent notation when writing equations.

This equation can be refined further to take account of Born forces. These are the repulsive electron–electron and nucleus–nucleus interactions between adjacent ions. Including the Born exponent, n, to account for these forces results in the Born–Landé equation:

$$\Delta U = -\left(N_A \times A \times \frac{|z_+||z_-|e^2}{4\pi\varepsilon_0 r}\right) \times \left(1 - \frac{1}{n}\right)$$

The value of n can be obtained from the electronic configurations of the constituent ions in the lattice (see Table 4.5). If the two ions have the same value of n, this is the value of n that should be used in the Born–Landé equation. If the values are different for both cation and anion, the average value is used.

Calculating the lattice enthalpy using the Born–Landé equation gives a value that can be compared with the experimentally derived value obtained from the Born–Haber cycle. Significant disagreement between the two values is usually an indication that the real lattice has a degree of covalent bonding character (i.e. the bonding cannot be fully described by a purely electrostatic or ionic model). This can occur when one of the ions is polarized by the oppositely charged ion (usually the anion is polarized by the cation) meaning that the electron cloud is distorted and the ion is no longer spherical.

Table 4.5 Values of the Born exponent, n, for various ions.

Electron configuration of ion	Example of cation	Example of anion	n
[He]	H^+, Li^+, Be^{2+}	H^-	5
[Ne]	Na^+, Mg^{2+}, Al^{3+}	F^-, O^{2-}, N^{3-}	7
[Ar] or [$3d^{10}$][Ar]	K^+, Ca^{2+}, Zn^{2+}	Cl^-, S^{2-}	9
[Kr] or [$4d^{10}$][Kr]	Rb^+, Sr^{2+}, Cd^{2+}	Br^-, Se^{2-}	10
[Xe] or [$5d^{10}$][Xe]	Cs^+, Ba^{2+}, Hg^{2+}	I^-, Te^{2-}	12

Source: Table 5.9 of Burrows, A., Holman, J., Parsons, A., Pilling, G., and Price, G. (2013) Chemistry[3], 2nd edn (Oxford University Press, Oxford).

Worked example 4.5A

Use the data in the table below and the Born–Landé equation to determine a theoretical value for the lattice enthalpy of potassium chloride. Compare the value you obtain with the value determined experimentally from a Born–Haber cycle: –690 kJ mol^{-1}.

Quantity	Value
r	317 pm
A	1.748
e	1.602×10^{-19} C
ε_0	8.854×10^{-12} C^2 J^{-1} m^{-1}

Solution

The Born–Landé equation is:

$$\Delta U = -\left(N_A \times A \times \frac{|z_+||z_-|e^2}{4\pi\varepsilon_0 r}\right) \times \left(1 - \frac{1}{n}\right)$$

We have values for most of the variables apart from n, the Born exponent. Using Table 4.5 we can see that $n = 9$ for K^+ and $n = 9$ for Cl^-. So the value of n for KCl = 9.

Substituting the values into the equation, and taking care with units, we get:

The units have been left in the first stage of the calculation to show how they cancel and leave a unit of J mol^{-1}, which can be divided by 1000 to give kJ mol^{-1}.

$$\Delta U = -\left(6.022 \times 10^{23}\ \text{mol}^{-1} \times 1.748 \times \frac{1 \times 1 \times (1.602 \times 10^{-19}\ \text{C})^2}{4 \times 3.142 \times 8.854 \times 10^{-12}\ \text{C}^2\ \text{J}^{-1}\ \text{m}^{-1} \times 317 \times 10^{-12}\ \text{m}}\right) \times \left(1 - \frac{1}{9}\right)$$

$$= -6.808 \times 10^5\ \frac{\text{mol}^{-1}\ \text{C}^2}{\text{C}^2\ \text{J}^{-1}\ \text{m}^{-1}\ \text{m}} = -681\ \text{kJ mol}^{-1}$$

The experimental value from the Born–Haber calculation is given as –690 kJ mol^{-1}, so the values are not too dissimilar. This is typically the case for alkali metal halides where the bonding is largely ionic.

Worked example 4.5B

Using the data in the table below and the Born–Landé equation, calculate the lattice enthalpy of CaF_2, giving your answer in kJ mol^{-1}.

Quantity	Value
$r(Ca^{2+})$	114 pm
$r(F^-)$	119 pm
A	2.519
E	1.602×10^{-19} C
$4\pi\varepsilon_0$	1.113×10^{-10} C^2 J^{-1} m^{-1}

Solution

The Born–Landé equation is:

$$\Delta U = -\left(N_A \times A \times \frac{|z_+||z_-|e^2}{4\pi\varepsilon_0 r}\right) \times \left(1 - \frac{1}{n}\right)$$

We are given the value of the Madelung constant, A, for CaF_2 as 2.519. We are also given the value of $4\pi\varepsilon_0$ as 1.113×10^{-10} C^2 J^{-1}m^{-1}. We need to determine the Born exponent, n, from the values in Table 4.4. For Ca^{2+} the Born exponent is 9 and for F^- the value is 7, so the value for CaF_2 is $(9+7)/2 = 8$.

For Ca^{2+} $|z_+|$ is 2 and for F^- $|z_-|$ is 1.

In this question we are given the ionic radii for Ca^{2+} and F^- and so must sum them to get the interionic distance: $r = 114$ pm + 119 pm = 233 pm.

Substituting the values in the equation we obtain:

$$\Delta U = -\left(6.022 \times 10^{23}\,\text{mol}^{-1} \times 2.519 \times \frac{2 \times 1 \times (1.602 \times 10^{-19}\,\text{C})^2}{1.113 \times 10^{-10}\,\text{C}^2\,\text{J}^{-1}\,\text{m}^{-1} \times 233 \times 10^{-12}\,\text{m}}\right) \times \left(1 - \frac{1}{8}\right)$$

$$= -2.627 \times 10^6\ \frac{\text{C}^2}{\text{C}^2\,\text{J}^{-1}\,\text{m}^{-1}}\frac{\text{mol}^{-1}}{\text{m}} = -2627\ \text{kJ mol}^{-1}$$

Question 4.9

The Madelung constant for NaCl is 1.74756 and the lattice enthalpy is -771 kJ mol^{-1}. Use the Born–Landé equation to determine the interatomic separation, r, in units of picometres.

Constants: $N_A = 6.022 \times 10^{23}$ mol^{-1}, $e = 1.602 \times 10^{-19}$ C, $\varepsilon_0 = 8.854 \times 10^{-12}$ C^2 J^{-1}m^{-1}.

Question 4.10

ZnO adopts the wurtzite structure, and has an experimentally measured lattice enthalpy of -4003 kJ mol^{-1}.

(a) Describe the structure of wurtzite.

(b) Given the thermodynamic data below, construct a Born–Haber cycle and use it to calculate the enthalpy of formation of ZnO.

Quantity	Physical process	Value/kJ mol^{-1}
$\Delta_{at}H^{\ominus}(\text{Zn})$	$\text{Zn(s)} \rightarrow \text{Zn(g)}$	130
$\Delta_{at}H^{\ominus}(\text{O})$	$\frac{1}{2}\text{O}_2\text{(g)} \rightarrow \text{O(g)}$	248
$IE_1(\text{Zn})$	$\text{Zn(g)} \rightarrow \text{Zn}^+\text{(g)} + \text{e}^-$	906
$IE_2(\text{Zn})$	$\text{Zn}^+\text{(g)} \rightarrow \text{Zn}^{2+}\text{(g)} + \text{e}^-$	1733
$\Delta_{EA1}H^{\ominus}(\text{O})$	$\text{O(g)} + \text{e}^-\text{(g)} \rightarrow \text{O}^-\text{(g)}$	−141
$\Delta_{EA2}H^{\ominus}(\text{O})$	$\text{O}^-\text{(g)} + \text{e}^-\text{(g)} \rightarrow \text{O}^{2-}\text{(g)}$	+844

(c) Use the Born–Landé equation to calculate the lattice enthalpy in kJ mol^{-1} for the wurtzite form of ZnO. The Madelung constant is 1.6381 and tabulated values of ionic radii are 74 pm for Zn^{2+} (four-coordinate) and 88 pm for Zn^{2+} (six-coordinate) and 126 pm for O^{2-}. Compare your answer to the experimentally derived value of -4003 kJ mol^{-1} and provide reasons for any discrepancy.

Values of ionic radii are difficult to specify absolutely as they depend upon the coordination geometry around the ion. A tetrahedral hole is smaller than an octahedral hole.

The Kapustinskii equation

The Born–Landé equation predicts reasonably accurate values for the lattice enthalpy for purely ionic solids, but its main drawback is that it relies on a knowledge of the structure of the solid so that values for the Madelung constant, A, and the interionic distances must be predicted or known. A simplification of the Born–Landé equation is the Kapustinskii equation which uses the fact that the ratio of the Madelung constant to the number of ions in the formula unit, ν, is roughly a constant. When this ratio is combined with the other constants in the Born–Landé equation, along with an average value for the Born exponent, a simplification of the Born–Landé equation is obtained:

The variable ν (pronounced 'nu') is the total number of ions in the formula unit.

$$\Delta U = -\frac{\kappa \nu |z_+||z_-|}{r_+ + r_-}$$

This is the Kapustinskii equation. The parameters $|z_+|$ and $|z_-|$ are the magnitudes of the charges on the ions and r_+ and r_- are the cation and anion radii in pm. The value v is equal to the total number of ions in the formula unit and κ is the Kapustinskii constant which has the value 107900 pm kJ mol^{-1}. The benefit of this equation over the Born–Landé equation is that it relies only on one constant which is not dependent upon the nature of the structure. In addition, the units are far simpler and give an energy in kJ mol^{-1}.

Worked example 4.5C

Use the Kapustinskii equation to obtain a value for the lattice enthalpy of potassium chloride, using ionic radii as given in Table 4.4. The Kapustinskii constant, κ, is 107 900 pm kJ mol^{-1}.

Solution

The Kapustinskii equation is:

$$\Delta U = -\frac{\kappa v |z_+||z_-|}{r_+ + r_-}$$

A common error when using the Kapustinskii equation is to multiply the ionic radii rather than sum them. Clearly if this mistake is made the units do not cancel to give an answer in kJ mol^{-1} and the overall lattice enthalpy value will be too small.

For KCl, v is 2, the radius of K^+ is 152 pm, and the radius of Cl^- is 167 pm. Inserting the values into the equation for lattice energy we obtain:

$$\Delta U = -\frac{107900 \text{ pm kJ mol}^{-1} \times 2 \times 1 \times 1}{152 \text{ pm} + 167 \text{ pm}} = -676 \text{ kJ mol}^{-1}$$

If we compare the value obtained here of –676 kJ mol^{-1} with that obtained in worked example 4.5A (–681 kJ mol^{-1}) we can see that the difference is small.

Worked example 4.5D

Determine the lattice energy of formation of calcium fluoride and the Kapustinskii equation using ionic radii as given in Table 4.4. The Kapustinskii constant, κ, has a value of 107 900 pm kJ mol^{-1}.

Solution

The Kapustinskii equation is:

$$\Delta U = -\frac{\kappa v |z_+||z_-|}{r_+ + r_-}$$

Comparing the value obtained here for CaF_2 with that obtained in Worked example 4.5C for KCl, we can immediately see the impact of increasing the charge of one of the ions (from +1 to +2) and decreasing the radii of the ions on the lattice enthalpy.

For CaF_2, v is 3, the radius of Ca^{2+} is 114 pm, and the radius of F^- is 119 pm. Inserting the values into the equation for lattice enthalpy we obtain:

$$\Delta U = -\frac{107900 \text{ pm kJ mol}^{-1} \times 3 \times 2 \times 1}{114 \text{ pm} + 119 \text{ pm}} = -2779 \text{ kJ mol}^{-1}$$

Worked example 4.5E

Calculate a value for the lattice enthalpy of CsBr using the Kapustinskii equation, and the following data:

Quantity	Value
Ionic radius of Cs^+	181 pm
Ionic radius of Br^-	182 pm
Kapustinskii constant, κ	107 900 pm kJ mol^{-1}

Compare your calculated value with that derived from the experimentally determined value of −632 kJ mol^{-1} obtained from a Born–Haber cycle calculation. See Worked example 4.4A.

Solution

The Kapustinskii equation provides a theoretical way to calculate the lattice enthalpy.

$$\Delta U = -\left(\frac{\kappa \nu |z_+||z_-|}{r_+ + r_-}\right)$$

- κ is the Kapustinskii constant, which has a value of 107 900 pm kJ mol^{-1}.
- ν is the number of ions in a single formula unit for the salt.
- $|z_+|$ and $|z_-|$ are the magnitudes of the charges on the cation and anion respectively.
- r_+ and r_- are the ionic radii in picometres for the cation and anion respectively.

$$\Delta U = -\left(\frac{107900 \times 2 \times |+1| \times |-1|}{181 + 182}\right) = -\left(\frac{107900 \times 2 \times 1 \times 1}{181 + 182}\right) = -594 \text{ kJ mol}^{-1}$$

The values obtained from the experimental data, via the Born–Haber cycle, and the value obtained by the Kapustinskii equation are relatively close together, differing by 38 kJ mol^{-1}. However, this is still a significant difference, with the experimentally determined lattice enthalpy being more exothermic. This implies that the bonding in CsBr is stronger than predicted by the Kapustinskii equation.

The Kapustinskii equation assumes that the ions are hard elastic spheres, and that the bonding is purely ionic in nature. These assumptions do not necessarily reflect the true nature of the ions or take into account other bonding effects, such as covalency. We should therefore not be surprised to see a discrepancy for a structure comprised of 'soft' or polarizable ions such as Cs^+ and Br^-.

Question 4.11

Use the Born–Landé equation and the Kapustinskii equation to obtain lattice enthalpies of formation for AgCl and compare with the experimentally determined value of −915 kJ mol^{-1} obtained from a Born–Haber cycle.

For AgCl: $A = 1.7476$; assume $r = \Sigma(r_+ + r_-)$ and obtain ionic radii values from Table 4.4 and values for the Born exponent, n, from Table 4.5.

$e = 1.602 \times 10^{-19}$ C, $\varepsilon_0 = 8.854 \times 10^{-12}$ $C^2 J^{-1} m^{-1}$, $\kappa = 107900$ pm kJ mol^{-1}.

Question 4.12

The following table gives the lattice enthalpies of a number of silver halides calculated either theoretically using the Kapustinskii equation or using experimentally determined values and Born–Haber cycles. Explain why differences between the two values increase on moving from AgF to AgI.

Material	Experimental (kJ mol^{-1})	Theoretical (kJ mol^{-1})	Difference (kJ mol^{-1})
AgF	−967	−953	14
AgCl	−915	−864	51
AgBr	−904	−830	74
AgI	−889	−808	81

Question 4.13

Use a Born–Haber cycle to find the enthalpy for formation of caesium nitride, Cs_3N, given the following thermochemical data. Comment on the result you obtain.

Quantity	Physical process	Value/kJ mol^{-1}
$\Delta_{at}H^{\ominus}(Cs)$	$Cs(s) \rightarrow Cs(g)$	+78.7
$IE_1(Cs)$	$Cs(s) \rightarrow Cs^+(g) + e^-$	+376
$\Delta_{at}H^{\ominus}(N)$	$\frac{1}{2}N_2(g) \rightarrow N(g)$	+473
$\Delta_{EA1}H^{\ominus}(N) + \Delta_{EA2}H^{\ominus}(N) + \Delta_{EA3}H^{\ominus}(N)$	$N(g) + e^- \rightarrow N^-(g)$ $N^-(g) + e^- \rightarrow N^{2-}(g)$ $N^{2-}(g) + e^- \rightarrow N^{3-}(g)$	+2565

Question 4.14

Use a Born–Haber cycle to find the enthalpy change for formation of magnesium nitride, Mg_3N_2, given the following thermochemical data: $r(Mg^{2+}) = 65$ pm, $r(N^{3-}) = 171$ pm.

Quantity	Physical process	Value/kJ mol^{-1}
$\Delta_{at}H^{\ominus}(Mg)$	$Mg(s) \rightarrow Mg(g)$	+150
$IE_1(Mg) + IE_2(Mg)$	$Mg(g) \rightarrow Mg^+(g) + e^-$ $Mg^+(g) \rightarrow Mg^{2+}(g) + e^-$	+2186
$\Delta_{at}H^{\ominus}(N)$	$\frac{1}{2}N_2(g) \rightarrow N(g)$	+473
$\Delta_{EA1}H^{\ominus}(N) + \Delta_{EA2}H^{\ominus}(N) + \Delta_{EA3}H^{\ominus}(N)$	$N(g) + e^- \rightarrow N^-(g)$ $N^-(g) + e^- \rightarrow N^{2-}(g)$ $N^{2-}(g) + e^- \rightarrow N^{3-}(g)$	+2565

Turn to the Synoptic questions section on page 148 to attempt questions that encourage you to draw on concepts and problem-solving strategies from several topics within a given chapter to come to a final answer.

Final answers to numerical questions appear at the end of the book, and fully worked solutions appear on the book's website. Go to http://www.oxfordtextbooks.co.uk/orc/chemworkbooks/.

References

Burrows, A., Holman, J., Parsons, A., Pilling, G., and Price, G. (2013) *Chemistry*^{3}, 2nd edn (Oxford University Press, Oxford).

Shannon, R. D. (1976) 'Revised Effective Ionic Radii and Systematic Studies of Interatomic Distances in Halides and Chalcogenides', *Acta Cryst.*, A32, 751–767.

Smart, L.E. and Moore, E. A. (2012) *Solid State Chemistry, An Introduction*, 4th edn (CRC Press, Boca Raton).

Lide, D. R. (ed.), (2006-7). *CRC Handbook of Chemistry and Physics*, 87th edn. CRC Press, Boca Raton, Florida.

5

Coordination complexes of the *d*-block metals

5.1 Occurrence of the *d*-block elements and the shapes of *d* orbitals

The *d*-block elements are the elements from Group 3 to Group 12 in the periodic table.

A transition element is defined as 'an element whose atom has an incomplete *d* subshell, or which can give rise to cations with an incomplete *d* subshell.' Thus the elements in Group 12, Zn, Cd, and Hg are not technically transition elements as they don't have any elements or ions with an incomplete *d* subshell as we shall see. It is important to be aware of the distinction between the two terms.

In this chapter we will use the Royal Society of Chemistry numbering system for groups in the periodic table, where the groups are numbered from 1 (alkali metals) to 18 (noble gases).

This block of elements is distinguished by having electrons in *d* orbitals. There are five *d* orbitals and it is important to be able to draw these orbitals and be able to label them according to their shape and orientation.

Worked example 5.1A

Sketch and name the set of *d* orbitals showing clearly the orientation of the boundary surfaces or lobes of the orbitals in each case.

The boundary surface of an orbital represents the volume in which there is about 90% probability of finding an electron at any given time.

Solution

The basic shape of four of the *d* orbitals is the same and consists of two dumb-bells at 90° to each other as shown here.

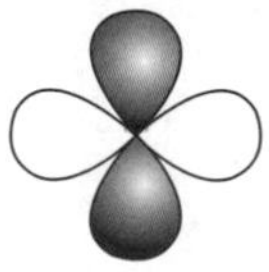

The shaded areas represent the phases of the orbital. Note that opposite lobes are in phase with each other and there are two planar nodes.

For three of the *d* orbitals the lobes point between the Cartesian axes. These are the d_{xy}, d_{xz}, and d_{yz} orbitals. One further *d* orbital has the same shape but its lobes point along the axes rather than between them. This is the $d_{x^2-y^2}$ orbital. The fifth *d* orbital, d_{z^2} has a completely different shape, although it has lobes which have components along all three axes. The shapes of the five *d* orbitals are shown in Figure 5.1.

A node is a region in an orbital where there is no possibility of finding an electron—see Chapter 1.

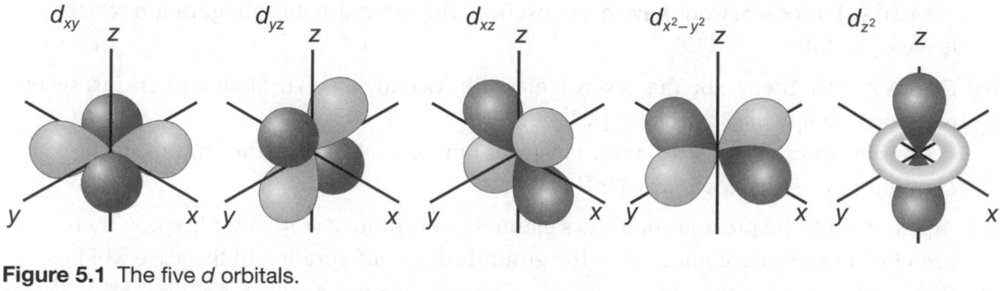

Figure 5.1 The five *d* orbitals.

5.2 Electronic configurations of the *d*-block metals and their ions

The symbol [Ar] means that the element has the same electronic configuration as argon for its first 18 electrons.

If we take the first row *d*-block elements (and we will be focusing on this row and its complexes) the first member is scandium, Sc (Z = 21), and the electronic configuration of scandium is : [Ar]$4s^2 3d^1$.

This is called the **ground state electronic configuration of the element**. As we go along the row the atomic number is increasing and an extra electron is placed in the *d* orbitals. At chromium, Cr (Z = 24), there is an associated stability with having a half-filled subshell of *d* electrons and so the ground state configuration is [Ar]$4s^1 3d^5$ rather than [Ar]$4s^2 3d^4$. A similar phenomenon is seen at copper, Cu (Z = 29), where it is more favourable to have the arrangement [Ar]$4s^1 3d^{10}$ with a filled set of *d* orbitals.

However, we aren't often required to refer to the ground state electronic configuration of a *d*-block element as most of the chemistry of these elements involves their complexes. The 4*s* orbital penetrates the core electrons far more than the 3*d* orbitals. Thus the 4*s* orbital experiences a higher effective nuclear charge and is therefore at a lower energy than the 3*d* orbitals and is filled first. On complexation the *d*-block elements lose their 4*s* electrons first, as these are now higher in energy than the 3*d* electrons, so all transition metal complexes will have their valence electrons in *d* orbitals, not *s* orbitals. This is very important to remember. Being able to determine the number of valence electrons of a metal in a complex is fundamental to understanding the properties of the complex. For example, iron (Fe) has the ground state configuration [Ar]$4s^2 3d^6$. The Fe(II) ion, Fe^{2+}, has the electronic configuration [Ar]$3d^6$. The two electrons have been lost from the 4*s* orbital.

The number of electrons lost is the same as the oxidation state of the metal.

Another point to note is that by numbering the groups in the periodic table horizontally from 1 to 18 the group number of the element gives us the total number of valence electrons for that transition element. So when a *d*-block element loses electrons we can simply subtract the number of electrons lost from the group number to get the number of remaining valence electrons.

The remaining electrons must all be in *d* orbitals. By following this rule it makes determining electronic configurations relatively straightforward.

Worked example 5.2A

Give the ground state electronic configurations and the electronic configurations for the metals in the following atoms and ions:

(a) Metallic titanium, Ti(s) and Ti^{3+}(g).

(b) Metallic cobalt, Co(s) and Co^{3+}(g).

(c) Metallic silver, Ag(s), and Ag^+(g).

Solution

Check! Ti is in Group 4 so it has four valence electrons. Losing three valence electrons leaves just one electron which is in the 3*d* orbital.

(a) Ti has Z = 22. After argon the next two electrons occupy the 4*s* orbitals and the last two electrons occupy the 3*d* orbital: [Ar]$4s^2 3d^2$.

► **Hint** Check the total number of electrons is 22; [Ar] = 18 + 4 = 22.

In Ti^{3+}(g) three electrons have been lost from the ground state configuration which leaves: [Ar]$3d^1$.

Check the total number of electrons is 27.

Check! Co is in Group 9 so it has nine valence electrons. Losing three valence electrons leaves six electrons which are in the *d* orbitals.

(b) Co has Z = 27. After argon the next two electrons occupy the 4*s* orbitals and the last seven electrons occupy the 3*d* orbitals: [Ar]$4s^2 3d^7$.

Co is in Group 9. In Co^{3+}(g) three electrons have been lost from the ground state configuration which leaves: [Ar]$3d^6$.

It is more favourable to have a filled set of *d* orbitals, [Kr]$5s^1 4d^{10}$ rather than a filled 5*s* orbital.

Check! Ag is in Group 11 so it has 11 valence electrons. Losing one valence electron leaves ten electrons which are in the *d* orbitals.

(c) Ag has Z = 47. The previous noble gas element is krypton, Z = 36. After krypton there are 11 electrons to accommodate. The ground state configuration in this case will be: [Kr]$5s^1 4d^{10}$.

In Ag^+(g) one electron has been lost from the ground state configuration which leaves: [Kr]$4d^{10}$.

Question 5.1

Give the ground state electronic configurations and the electronic configurations for the metals in the following atoms and ions:

(a) Metallic nickel, Ni(s) and Ni^{2+}(g).

(b) Metallic manganese, Mn(s) and Mn^{7+}(g).

(c) Metallic tungsten, W(s) and W^{3+}(g).

In this workbook we will refer to compounds formed by the transition metals and ligands as 'coordination complexes'. They are sometimes called 'coordination compounds'.

5.3 Determining oxidation states of *d*-block metal complexes and their electronic configurations

If we have a transition metal complex the oxidation state of the metal is determined in exactly the same way as for a main group complex. We can assign an oxidation state to the metal even if the complex is not purely ionic. It's useful to be able to assign an oxidation state in order to work out the electronic configuration. For uncharged binary complexes the charge on the metal will balance the charge on the associated counter ions or ligands.

A ligand is an atom or group of atoms that undergoes chemical bonding to the central metal ion.

For example, in $FeCl_3$, three singly negatively charged chloride ions provide a total negative charge of −3. The metal must therefore be +3. If the complex is charged (a complex ion) we subtract the charge of the ion before calculating the charge on the metal. For example, in $[FeCl_4]^{2-}$ the chloride ions provide an overall negative charge of 4–. Two of these electrons provide the 2– charge on the complex ion leaving a charge of −2 to be balanced by the metal. So the metal is in the +2 oxidation state.

Remember: charge of the metal ion + charge of the ligands = charge on the complex.

This can be shown mathematically by assigning the oxidation state of the metal to be x. We can construct the equation: $x + (-4) = -2$. So $x = -2 - (-4) = +2$.

Being able to determine the oxidation state of the metal depends upon having a knowledge of the charge on the associated counter ions or ligands. It is useful to be familiar with common ligands and their charges and these can be found in most standard text books. Many ligands such as H_2O, NH_3, CO, etc. are uncharged. These do not affect the oxidation state of the metal. We can consider them as bonding through the lone pair of electrons on the donor atom.

A donor atom is the atom in the ligand that bonds directly to the metal.

So in the complex chromium hexacarbonyl, $Cr(CO)_6$, the chromium atom is in the zero oxidation state and the CO ligand is not charged.

Worked example 5.3A

Determine the oxidation states and the electronic configurations of the metal ions in the following coordination complexes and complex ions:

(a) $[TiCl_4]^{2-}$

(b) $[Co(H_2O)_6]Cl_2$

(c) $MoOCl_3$

(d) $Ni(CO)_4$

Solution

(a) $[TiCl_4]^{2-}$

The chloride ions have a total charge of −4 and the complex ion has a charge of −2.

Let the oxidation state of the Ti = x, then $x + (-4) = -2$ so $x = -2 - (-4) = +2$.

The ion is Ti^{2+}. As Ti is in the first row of the *d*-block elements and in Group 4, the electronic configuration is $[Ar]3d^2$.

(b) $[Co(H_2O)_6]Cl_2$

The complex ion has an overall charge of +2 to balance the charge of the two chloride ions, so the formula is $[Co(H_2O)_6]^{2+}$. As the water molecules are uncharged the cobalt is therefore in the +2 oxidation state, Co^{2+}.

Cobalt is in Group 9, it has lost two electrons and so it has seven electrons remaining, all in *d* orbitals. So the electronic configuration is: $[Ar]3d^7$.

(c) $MoOCl_3$

This is an uncharged complex of molybdenum. The oxygen ligand can be assumed to be present as O^{2-} and the three chlorine ligands as Cl^-. Thus the total charge of the ligands is $-2 + 3 \times (-1) = -5$. The molybdenum ion can be assumed to have a charge of +5 (although the bonding is not ionic in this compound). Molybdenum is in Group 6, so has lost five electrons and has one electron remaining in the valence shell. The electronic configuration will therefore be: $[Kr]4d^1$.

a) $Ni(CO)_4$

The CO ligand is uncharged and so Ni is in the zero oxidation state and present as Ni(0).

The electronic configuration is not the same as the ground state configuration as the element has reacted with the carbon monoxide and formed a new complex. All valence electrons are now *d* electrons. Ni is in Group 10, so as it hasn't lost any valence electrons the electron configuration is: $[Ar]3d^{10}$.

Question 5.2

What is the oxidation state of the metal and the valence shell electronic configuration of the metal in each of the following complexes?

(a) $[Mn(H_2O)_6]^{2+}$
(b) $[NiCl_4]^{2-}$
(c) $[Cr(ox)_3]^{3-}$ ox = $(C_2O_4)^{2-}$
(d) $[Cr(CO)_6]$

5.4 Geometry and coordination number in coordination complexes

A Lewis acid is a species that can accept a pair of electrons and a Lewis base is a species that can donate a pair of electrons. See Chapter 3.

The **coordination number** of a metal is the number of donor atoms from ligands that are bonded to the metal centre. *d*-block metals act as Lewis acids by accepting a pair of electrons and the surrounding atoms or molecules that bond to the metal act as Lewis bases and are called ligands. The donor atom of the ligand donates a pair of electrons to the metal as in co-ordinate bonding. Ligands which have just one donor atom are called unidentate ligands, for example, **Cl**$^-$, H_2**O**, **N**H_3. Ligands with two donor atoms are called bidentate ligands, for example, ethane-1,2-diamine, en (**N**$H_2CH_2CH_2$**N**H_2); ethanedioate, ox^{2-} (CO**O**$^-$)$_2$. There are also tridentate (3), tetradentate (4), pentadentate (5), and hexadentate (6) ligands. Examples of each of these ligands are illustrated in Figure 5.2.

Thus for a complex such as $[Co(en)_2Cl_2]$ there is a total of six donor atoms and the coordination number is therefore six. The complex $[NiCl_2(PMe_3)_2]$ has a total of four donor atoms and the coordination number is four.

The **coordination geometry** is the shape that the molecule adopts. This is generally the shape that results when all the ligands repel each other so as to minimize steric and electronic repulsions. Complexes with a coordination number of six are usually octahedral. However, this rule doesn't always result in the correct geometry as there are also electronic effects to consider. There are two principal geometries for four-coordinate transition metal complexes: square planar and tetrahedral; certain metals with certain numbers of *d* electrons will preferentially adopt one or other

bidentate ligands

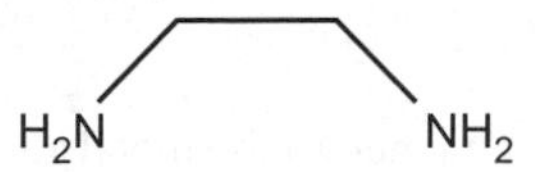

ethane-1,2-diamine or ethylene diamine en

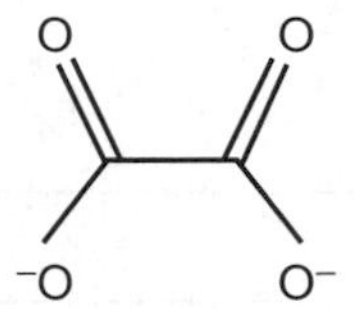

ethanedioate ion or oxalate ion ox^{2-}

tridentate ligands

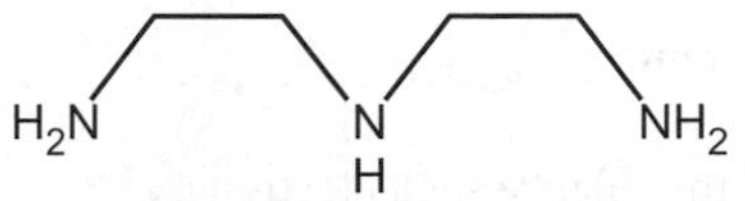

1,4,7-triazaheptane or diethylenetriamine dien

hexadentate ligands

diamino-1,2-ethane-*N,N,N',N'*-tetraethanoate ion

ethylenediaminetetraacetate ion $[edta]^{4-}$

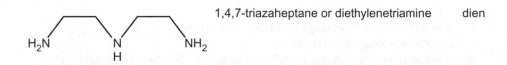

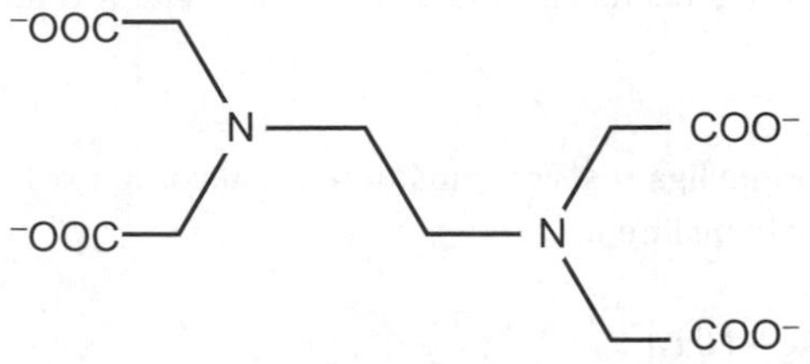

Figure 5.2 Examples of some bidentate, tridentate, and hexadentate ligands.

configuration. As we have seen for main group elements a variety of different configurations can be adopted for different coordination numbers.

VSEPR doesn't work as well to predict the shapes of *d*-block complexes as it does for *p*-block complexes. The shapes of transition metal coordination complexes are determined largely by the number of ligands and the steric and electronic properties of the ligands. The size (and therefore oxidation state) of the central metal ion is also important.

Worked example 5.4A

Give the coordination numbers and suggest the geometries of the metal ions in the following complexes:

(a) $[Ru(NH_3)_5(NCMe)]^{2+}$

(b) $[ReCl(CO)_3(py)_2]$ (py = C_5H_5N)

(c) $[CoCl_4]^{2-}$

Solution

(a) $[Ru(NH_3)_5(NCMe)]^{2+}$

Here there are six monodentate ligands attached to the central metal and so the coordination number is six. The geometry is likely to be octahedral.

(b) $[ReCl(CO)_3(py)_2]$

All six ligands in this complex are unidentate and so the coordination number is six and the complex is likely to be octahedral.

(c) $[CoCl_4]^{2-}$

In this complex there are four unidentate chloride ligands attached to the metal and so the coordination number is four. The complex is likely to be tetrahedral or square planar. In this case the complex adopts tetrahedral geometry which reduces both steric and electronic interactions.

Worked example 5.4B

For the following transition metal complexes :

(a) $[Ag(NH_3)_2]^+$

(b) $[Pt(en)Cl_2]$ (en = $NH_2CH_2CH_2NH_2$)

(c) $[Mo(CO)_6]$

i. Sketch the shape of the complex and name the geometry.
ii. Give the oxidation state of the metal.
iii. Give the number of *d* electrons on the metal and the valence shell electron configuration of the metal.

Solution

► **Hint** For all these complexes first determine the coordination number of the metal and then decide the most likely geometry.

(a) $[Ag(NH_3)_2]^+$

i. The Ag^+ ion is coordinated by two ammonia ligands. For transition metal complexes this generally means that the complex will be linear.

$$[H_3N{-}Ag{-}NH_3]^+$$

ii. The ligands are neutral ammonia molecules therefore the silver ion is in the +1 oxidation state.

iii. Silver is in Group 11 so the number of electrons is 11 – 1 = 10. As silver is in row 5, these are 4*d* electrons so the electronic configuration is [Kr] $4d^{10}$.

► **Hint** To determine the number of *d* electrons find the group that element is in and subtract the charge. This gives the number of *d* electrons.

How do we know it's square planar rather than tetrahedral? This is always a difficult question. Often you are given some information about the complex that can help. In this case you should use the fact that Pt is a third row transition element in Group 10. Such complexes are often square planar and the reason for this is the arrangement of the electrons in the *d* orbitals. You will cover this when you study crystal-field theory in some depth.

d-block elements towards the right-hand side often have oxidation states of +2 as the first two *s* electrons are lost readily.

Group 10 metals frequently have eight *d* electrons in their valence shell—this is a particularly stable configuration if square planar.

(b) $[Pt(en)Cl_2]$

i. en is a bidentate ligand and coordinates through the two N atoms and the Cl^- ligands are unidentate so the platinum ion is four coordinate. Four-coordinate complexes can generally be tetrahedral or square planar. In this case the complex is square planar:

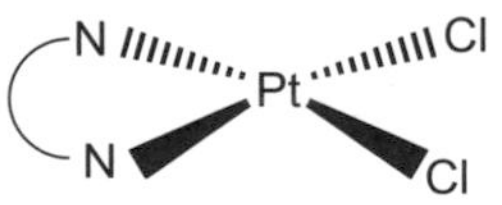

ii. The en ligand is uncharged and each chloride ligand has a charge of –1. This means that the oxidation state of the Pt is +2.

iii. Pt is in Group 10 so the number of *d* electrons is 10 – 2 = 8. As Pt is a third transition-row element the electron configuration is [Xe] $5d^8$.

(c) $[Mo(CO)_6]$

i. This complex has a central Mo atom surrounded by six CO (carbonyl) ligands. The most likely geometry is therefore octahedral.

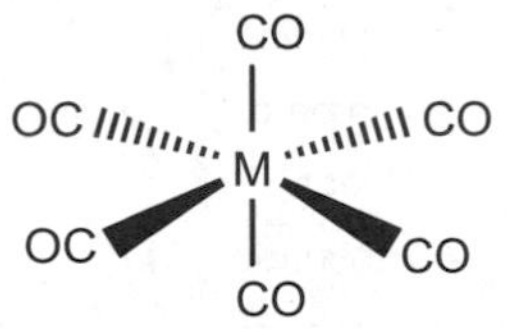

ii. CO is an uncharged ligand and so the Mo atom must be in the zero oxidation state.

iii. As the Mo atom is in the zero oxidation state no electrons have been lost. The metal is in Group 6 and so has a total of six valence electrons. These are all *d* electrons and as Mo is a second-row transition element these are 4*d* electrons so the electron configuration is [Kr] $4d^6$.

A common mistake when determining electronic configurations of metals in the zero oxidation state is to assume that because no electrons have been lost the electron configuration of the complexed metal atom is the same as in the ground state of the uncoordinated free atom. This is not the case as on complexation molecular orbitals of the metal interact with those of the ligand and the *d* orbitals are then lower in energy than the valence *s* orbital.

Question 5.3

Give the coordination numbers and suggest the geometries of the metal ions in the following complexes:

(a) $NH_4[Cr(NH_3)_2(NCS)_4].H_2O$

(b) $VOCl_3$

(c) $[Ni(edta)]^{2-}$

(d) $[Cr(en)_2F_2]^+$

5.5 Naming coordination complexes

When writing the name of a coordination complex the name is written as one word with the ligands before the metal and its oxidation state in brackets after the metal in Roman numerals.

If the complex ion is anionic the metal is given the ending **-ate**, for example chromium becomes chromate. In some cases the metal takes the Latin name, for example iron becomes ferrate and gold becomes aurate. See Table 5.1 for the names of common metal ions when in complex anions.

If the complex contains more than one type of ligand the names of the ligands are given alphabetically. When writing the formula for the complex ion the metal comes first, followed by the ligands. Anionic ligands are listed before neutral ligands. Usually the name of the ligand is the same as that of the free molecule, but there are exceptions such as aqua (water), ammine

Table 5.1 Nomenclature of metal atoms in complex anions.

Element	Symbol	Name in an anion
vanadium	V	vanadate
chromium	Cr	chromate
manganese	Mn	manganate
iron	Fe	ferrate
cobalt	Co	cobaltate
nickel	Ni	nickelate
copper	Cu	cuprate
zinc	Zn	zincate
silver	Ag	argentate
gold	Au	aurate

Table 5.2 Some common anionic ligands.

Name of ion	Formula	Charge	Abbreviation	Name of coordinated ligand	IUPAC (formal) name
fluoride	F^-	–1		fluoro	fluorido
chloride	Cl^-	–1		chloro	chlorido
bromide	Br^-	–1		bromo	bromido
iodide	I^-	–1		iodo	iodido
hydride	H^-	–1		hydride	hydrido
oxide	O^{2-}	–2		oxo	oxo
hydroxide	OH^-	–1		hydroxo	hydroxo
cyanide	CN^-	–1		cyano	cyanido
isocyanide	NC^-	–1		isocyano	isocyanido
thiocyanide	SCN^-	–1		thiocyano	thiocyanido
isothiocyanide	NCS^-	–1		isothiocyano	isothiocyanido
sulfate	SO_4^{2-}	–2		sulfato	sulfato
nitrite	NO_2^-	–1		nitro	nitrito-*N*
nitrite	$O\text{-}N\text{-}O^-$	–1		nitrito	nitrito-*O*
ethanedioate ion	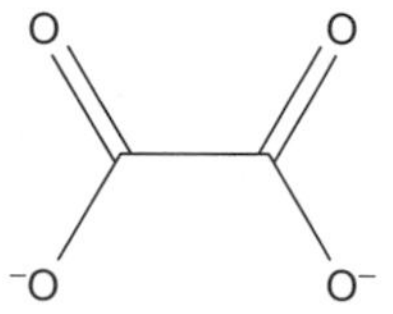	–2	$[ox]^{2-}$	oxalato	ethanedioato
glycine	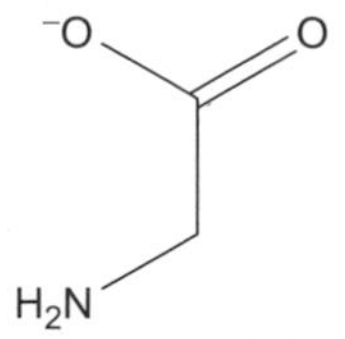	–1	$[gly]^-$	glycine	glycinato
acetylacetonate anion	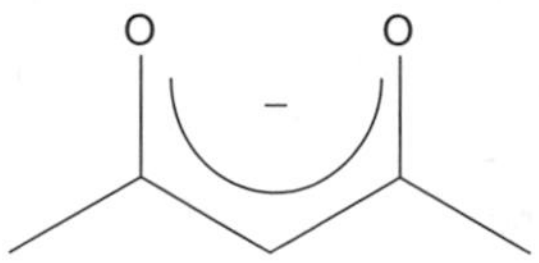	–1	$[acac]^-$	acetylacetonate	pentan-2,4-dionato
diamino-1,2-ethane-*N,N,N′,N′*-tetraethanoate ion	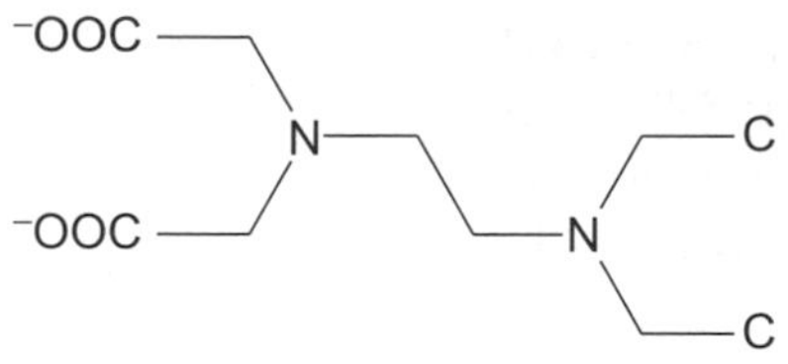	–4	$[edta]^{4-}$	ethylenediamine tetraacetate	diamino-1,2-ethane-*N,N,N′,N′*-tetraethanoato

The most recent IUPAC recommendation is that an anionic ligand is given the name of the free ligand with the ending changed from –e to –o. So chloride becomes chlorido. See IUPAC (1997). However, in common usage we often use 'chloro' in the name of the complex.

(ammonia), and carbonyl (carbon monoxide). Anionic ligands are given the ending -o, such as chloro, cyano, etc.

Table 5.2 lists some common anionic monodentate ligands, while Table 5.3 lists some common neutral ligands.

Ambidentate ligands and the kappa convention

Note that some ligands can coordinate through two different donor atoms. These are called **ambidentate** ligands. In Table 5.2, **C**N^- (cyano) and **N**C^- (isocyano), **S**CN^- (thiocyano) and **N**CS^- (isothiocyano), and **N**O_2^- (nitro) and **O**–N–O^- (nitrito) are ambidentate ligands. The atom they coordinate through is shown in bold in each case.

Table 5.3 Some common neutral ligands.

Name of molecule	Formula of molecule	Abbreviation	Name of coordinated ligand
water	$H_2\mathbf{O}$		aqua
ammonia	$\mathbf{N}H_3$		ammine
carbon monoxide	$\mathbf{C}O$		carbonyl
nitric oxide	$\mathbf{N}O$		nitrosyl
amines e.g. $MeNH_2$	$CH_3\mathbf{N}H_2$		methylamine
pyridine	$C_5H_5\mathbf{N}$		pyridine
trimethylphosphine	$\mathbf{P}Me_3$		trimethylphosphine
triphenylphosphine	$\mathbf{P}Ph_3$		triphenylphosphine
ethane-1,2-diamine	$\mathbf{N}H_2CH_2CH_2\mathbf{N}H_2$	en	ethane-1,2-diamine
bipyridine	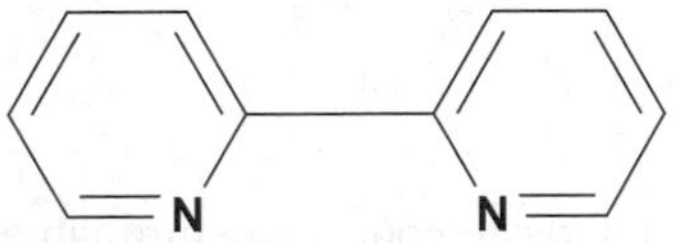	bipy	2,2′-bipyridine 2,2′-bipyridyl

Atoms in **bold** are the donor atoms that coordinate to the metal centre.

Note that, for the nitrite ligand, when NO_2^- is coordinated through N it takes the name nitro or nitrito-***N*** and when coordinated through the oxygen atom as O–N–O⁻ it takes the name nitrito or nitrito-***O***. However, to precisely indicate the donor atom the kappa convention is used. For example, if the NCS^- ligand is bonded through the N atom (as in isocyanato) the name of the complex would include 'thiocyanato-κ**N**' and if the ligand were bonded through sulfur the name would include 'thiocyanato-κ**S**.' The Greek letter kappa, κ, is placed in the name of the complex after the portion of the ligand name that represents the chain or ring or substituent group in which the ligating atom is found. In this way nitrogen bonded NO_2^- is named nitrito-κ**N** and oxygen bonded nitrito is named nitrito-κ**O**.

When two or more identical ligands or parts of a polydentate ligand are present, a superscript is used to the right of κ to indicate the number of such bonded atoms. So the complex $Pt[(CH_3)_2PCH_2CH_2P(CH_3)_2]Cl_2$ would take the name dichlorobis[1,2-ethanediylbis(dimethylphosphine)-κ^2**P**]platinum(II). In many cases this convention in naming complexes is ignored and trivial names are used, or the coordinated atom is indicated by using italics in the name of the complex. We will not be demonstrating examples of the kappa convention in the worked examples below. This convention is more useful for complex multidentate chelating ligands.

If there is more than one type of a certain ligand the prefixes di, tri-, tetra-, penta-, etc. are used to indicate the number. But if the ligand already contains a prefix in its name then the terms bis-, tris-, tetrakis-, etc. are used. The notation for representing multiple ligands is summarized in Table 5.4.

Table 5.4 Ligand multiplicity.

Number of ligands	Standard prefix	Prefix when ligand contains a numerical prefix
1	mono	
2	di	bis
3	tri	tris
4	tetra	tetrakis
5	penta	pentakis
6	hexa	hexakis

Hapticity of coordinated ligands

Certain ligands such as alkenes contain double bonds and each double bond coordinated to the metal contributes a pair of electrons. However, ligands with multiple double bonds can use one or more double bonds to coordinate to the metal centre. The hapticity of the ligand indicates the number of atoms from the ligand that are directly bonded to the metal. The hapticity is indicated by the Greek symbol eta, η, followed by a superscript. So ethene generally coordinates through the double bond and both carbon atoms are considered attached to the metal. In this case the ligand is said to be η^2 coordinated. For example $[PtCl_3(\eta^2\text{-}C_2H_4)]^-$ has the structure:

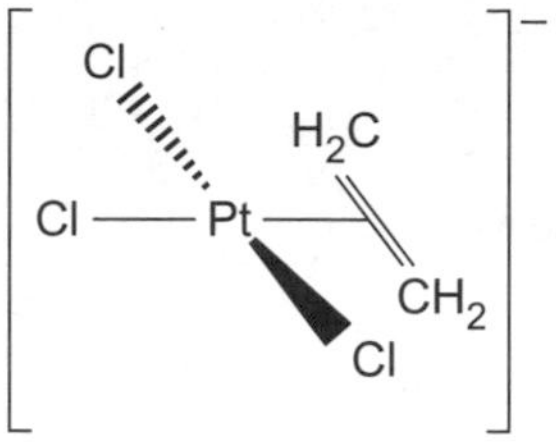

Ligands with more than one double bond can coordinate through any or all of the double bonds. In butadienetricarbonyliron(0) shown here, the butadiene is coordinated through both double bonds and so the formula is: $[Fe(CO)_3(\eta^4\text{-}C_4H_6)]$ and has this structure:

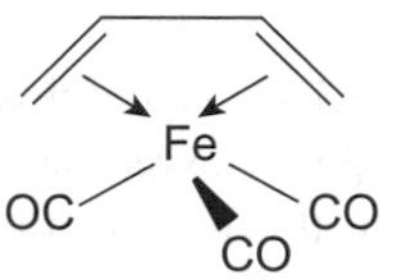

Worked example 5.5A

Give the name of the complex ion whose formula is $[CrCl_2(H_2O)_4]^+$.

Solution

The ligands here are Cl and H_2O, which are named 'chloro' and 'aqua.' 'Aqua' comes before 'chloro' alphabetically. There are four H_2O molecules and two chloro ligands and so we have tetraaqua and dichloro.

The metal is chromium and it is part of a complex cation. To work out the oxidation state of the metal we see the ion has an overall single positive charge. As there are two chloride ligands the metal must be in the +3 oxidation state, so chromium (III).

The overall name is therefore: tetraaquadichlorochromium(III) ion.

Worked example 5.5B

Give the name of the complex ion whose formula is $[Co(OH_2)(NH_3)(en)_2]^{3+}$.

Solution

Here we have three different ligands, one water molecule named aqua, one ammonia ligand named ammine, and two ethane-1,2-diamine molecules.

Thus the ammine comes first, then the aqua, and then the en ligands. There are two en ligands but as the prefix 'di' is found in their name they are numbered bis-ethane-1,2-diamine.

The ion has a three positive charge and as none of the ligands is charged the oxidation state of the metal is the same as that of the charge which is +3.

Thus the overall name is: amminemonoaquabis-ethane-1,2-diamlne cobalt(III).

Note that the ammonia molecule as a ligand has the coordinated name 'ammine' (with double m), whereas when amines are coordinated they take the name 'amine' (with just one m). This is true for bidentate amine ligands such as ethane-1,2-diamine or 'en'.

Worked example 5.5C

Give the name of the complex ion whose formula is $K_4[Fe(CN)_6]$.

Solution

This species contains a complex anion, $[Fe(CN)_6]^{4-}$. There is only one type of ligand, the cyano ligand and there are six of these so this is indicated by 'hexacyanido' or 'hexacyano'. The complex ion has a charge of –4 so the metal must be in the +2 oxidation state. As it is an anion we have to use the name ferrate. So the name is: potassium hexacyanoferrate(II).

Remember that iron is one of the metals that takes a Latin form, in this case 'ferrate' when in an anionic complex.

Question 5.4

Give the names of the following complexes:

(a) $[Cr(SCN)(NH_3)_5]^{2+}$

(b) $[Co(NO_2)_2(NH_3)_4]^+$

(c) $[Mn(ox)_2(H_2O)_2]^{2-}$

(d) $Fe(CO)_3(PPh_3)_2$

5.6 Writing the formulae of coordination complexes

The formula of a coordination complex is written in a different order than its name. The chemical symbol for the metal centre is written first. The ligands are written next, with anionic ligands coming before neutral ligands. If there is more than one anion or neutral ligand, they are written in alphabetical order according to the first letter in their chemical formula.

Worked example 5.6A

Give the formula of the complex whose name is pentaamminechlorocobalt(III) sulfate.

The IUPAC rules recommended the use of chlorido in place of chloro for Cl, but chloro is more commonly found.

Solution

In this coordination complex we have a complex cation and a sulfate anion. The formula of the cation precedes the anion. In the cation the metal is listed first, followed by the anionic ligands—in this case the one chloro ligand and then the five neutral ammonia ligands.

The formula of the complex is: $[CoCl(NH_3)_5]SO_4$.

Note that square brackets are commonly used to enclose the complex ion and depict the coordination sphere of the metal.

Worked example 5.6B

Give the formula of the complex whose name is potassium hexacyanochromate(III).

Solution

Here the complex ion is anionic (hexacyanochromate(III)), so the counter cations, K^+, precede the formula of the metal ion. The charge on the anion is –3 so there must be three potassium ions. There is only one type of ligand and so no choice about the order in the formula. The formula of the complex is: $K_3[Cr(CN)_6]$.

If the ligand contains more than one atom it is usually given in round brackets so there is no doubt over how many of that type of ligand are present.

Worked example 5.6C

Give the formula of the complex whose name is tetraamminedibromocobalt(III) chloride.

Solution

Here we have a complex cation containing the metal surrounded by two different ligands. The bromide (Br^-) ligand is anionic and so precedes the four ammonia ligands. The formula is therefore: $[CoBr_2(NH_3)_4]Cl$.

Question 5.5

Give the formulae for the following coordination complexes:

(a) diaquadibromodi(methylamine)chromium(III) nitrate.

(b) tricarbonylmonochloroglycinatoruthenium(0) (glycinato = $NH_2CH_2COO^-$).

(c) dichlorobis(ethane-1,2-diamine)cobalt(III) chloride.

5.7 Isomerism in coordination complexes

Two main types of isomerism can occur in coordination complexes as with molecules in organic chemistry. These fall under the headings of structural isomerism and stereoisomerism.

We will briefly look at the different types of structural isomerism first and then consider stereoisomerism.

Structural isomerism

Structural isomers are defined as complexes in which the same atoms are present but joined in a different way. Many different types of structural isomers are possible for coordination complexes.

Ionization and hydration isomers

These isomers result from the interchange of ligands within the coordination sphere of the metal with those outside of it. If the exchanging ligands are ions, we have **ionization isomers**, for example: $[PtCl_2(NH_3)_2](NO_2)_2$ and $[Pt(NO_2)_2(NH_3)_2]Cl_2$.

If the exchanging ligands are water molecules the isomers are called **hydration isomers**: $[CrCl(H_2O)_5]Cl_2$ and $[CrCl_2(H_2O)_4]Cl.H_2O$.

Coordination isomers

When a coordination complex contains both a complex cation and complex anion it is sometimes possible for the ligands to exchange between the metal centres, for example: $[Cr(NH_3)_6][Co(CN)_6]$ and $[Co(NH_3)_6][Cr(CN)_6]$.

Linkage isomers

These are isomers which have the same composition but contain one or more ligands that can attach to the metal through different donor atoms. Thus the coordination sphere of the metal differs in each isomer. These ligands are called ambidentate and examples are: **SCN^-** - thiocyanate and **NCS^-** - isothiocyanate; **NO_2^-** - nitro (or nitrito-*N*) and **ONO^-** - nitrito (or nitrito-*O*).

An example of such an isomer would be $[Co(\mathbf{O}NO)(NH_3)_5]Cl_2$ where the O is attached to the cobalt and $[Co(\mathbf{N}O_2)(NH_3)_5]Cl_2$ where the N is attached to the cobalt.

Worked example 5.7A

Identify the type of structural isomerism present in the following pairs of isomers of coordination complexes:

(a) $[CoBr(H_2O)_5]Cl$ and $[CoCl(H_2O)_5]Br$

(b) $[Co(NCS)(NH_3)_5]^{2+}$ and $[Co(SCN)(NH_3)_5]^{2+}$

(c) $[Zn(NH_3)_4][CuCl_4]$ and $[Cu(NH_3)_4][ZnCl_4]$

Solution

(a) $[CoBr(H_2O)_5]Cl$ and $[CoCl(H_2O)_5]Br$
In these complexes the chloride and bromide ligands have exchanged so that in the first complex the chloride ion acts as the counter ion whereas in the second complex the bromide ligand is the counter ion and the chloride ligand is in the coordination sphere of the metal. These are therefore **ionization** isomers.

(b) $[Co(NCS)(NH_3)_5]^{2+}$ and $[Co(SCN)(NH_3)_5]^{2+}$
In these two complexes there is a thiocyano ligand which is ambidentate. Thus these are **linkage** isomers.

(c) $[Zn(NH_3)_4][CuCl_4]$ and $[Cu(NH_3)_4][ZnCl_4]$
In these two complexes the ligands in the coordination spheres of the metals have exchanged. In the first complex the zinc is tetrahedrally surrounded by ammonia molecules whereas in the second complex it is now surrounded by chloride ligands. Thus these are **coordination** isomers.

Question 5.6

(a) Give a possible coordination isomer of $[Fe(CN)_2(bipy)_2][Co(bipy)(CN)_4]$.

(b) Give a possible ionization isomer of $[PtBr(NH_3)_3]NO_2$.

(c) Give a possible linkage isomer of $[CoCl(NO_2)(NH_3)_4]Cl$.

(d) Predict the linkage manner of the thiocyano ligand in the complex thiocyanatotrimethylphosphinegold(I).

Stereoisomerism

Stereoisomers have the same atoms bound in the same way but with different positions of the atoms in space. Stereoisomers can be **geometric** isomers which are possible for square planar and octahedral species and **optical** isomers which are possible for tetrahedral and octahedral complexes, but not square planar.

Geometric isomers

The simplest type of geometric isomer is when the same complex can adopt different geometries. For example $[Ni(CN)_5]^{3-}$ can be either square pyramidal or trigonal bipyramidal depending upon the cation present (Figure 5.3).

Four-coordinate complexes of transition metals are either tetrahedral or square planar in geometry (Figure 5.4), and the geometry adopted depends upon a number of factors.

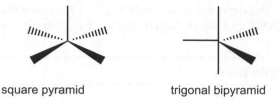

Figure 5.3 Square pyramidal and trigonal bipyramidal coordination geometries.

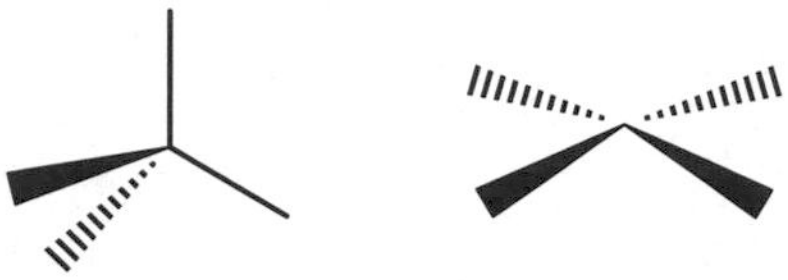

Figure 5.4 Tetrahedral and square planar coordination geometries.

In square planar complexes *cis* and *trans* isomerism is possible. The *cis* isomer has the same ligands adjacent to each other whereas the *trans* isomer has the same ligands opposite to each other across the metal centre. The best known example of *cis/trans* isomerism is the anti-cancer agent *cis*-platin (*cis*-diamminedichloroplatinum(II)), shown in Figure 5.5.

Some octahedral complexes with general formula MA_2B_2 (where A and B are different unidentate ligands) can also have *cis* and *trans* isomers—for example, $[CoCl_2(NH_3)_4]^+$ as shown in Figure 5.6.

Octahedral coordination complexes with general formula MA_3B_3 can also have *fac/mer* isomerism. In the *fac* isomer three of the same ligands occupy one face of the octahedron whereas in the *mer* isomer equivalent ligands occupy the meridian, as shown in Figure 5.7 for $[CoCl_3(CN)_3]^{3-}$.

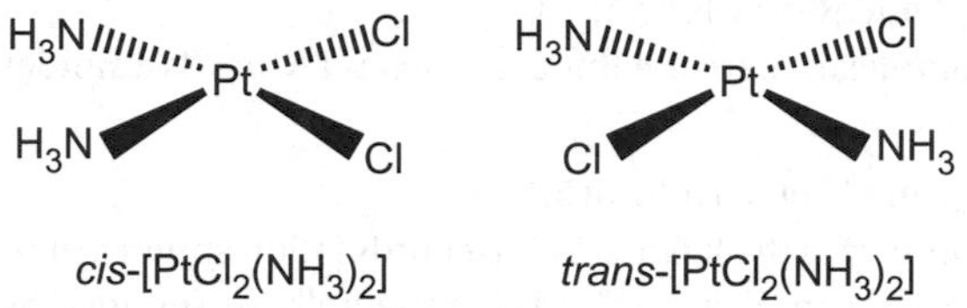

Figure 5.5 *cis*- and *trans*-diamminedichloroplatinum(II).

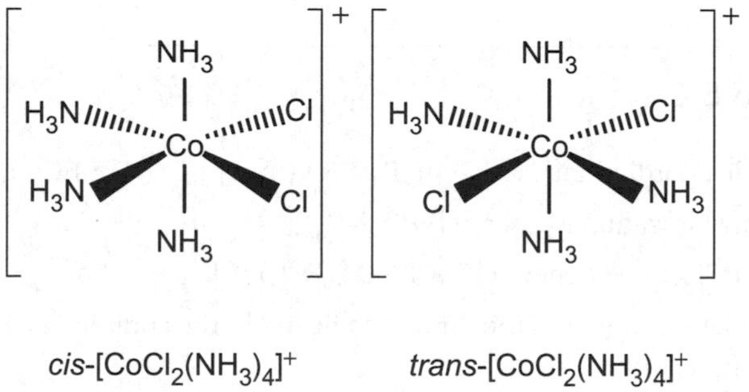

Figure 5.6 *cis*- and *trans*-tetraamminedichlorocobalt(III) cation.

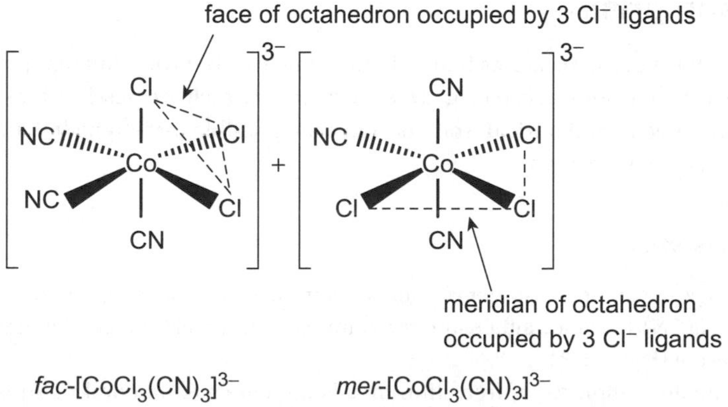

Figure 5.7 *fac* and *mer* isomerism in the trichlorotricyanidocobaltate(III) ion.

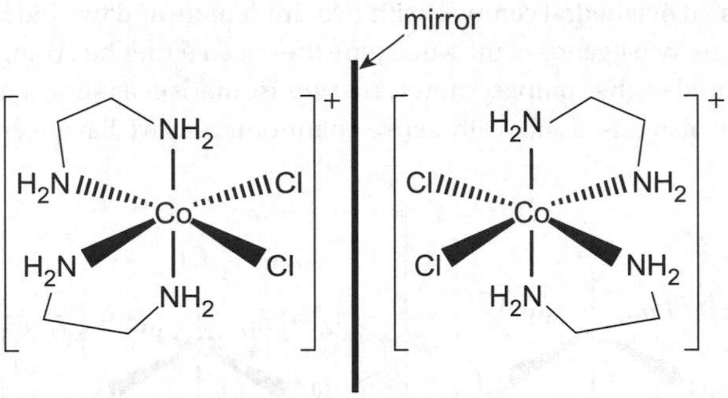

Figure 5.8 Enantiomers of $[CoCl_2(en)_2]^+$.

Optical isomers

Optical isomers are non-superimposable mirror images and differ in the direction in which they rotate the plane of polarized light. Such isomers are called enantiomers. Both tetrahedral and octahedral coordination complexes can show optical isomerism, and common examples in coordination chemistry are found with complexes that contain mutually cis-bidentate ligands, such as in $[CoCl_2(en)_2]^+$ shown in Figure 5.8.

Worked example 5.7B

State which of the following coordination complexes have stereoisomers and sketch two possible isomers.

(a) $[Co(NO_2)_3(NH_3)_3]$

(b) $[CoCl_2(en)_2]^+$

(c) $Fe(en)_3$

Solution

(a) $[Co(NO_2)_3(NH_3)_3]$ is an octahedral complex with an MA_3B_3 arrangement of ligands. It can therefore have *fac*/*mer* isomerism. Possible isomers are shown in Figure 5.9.

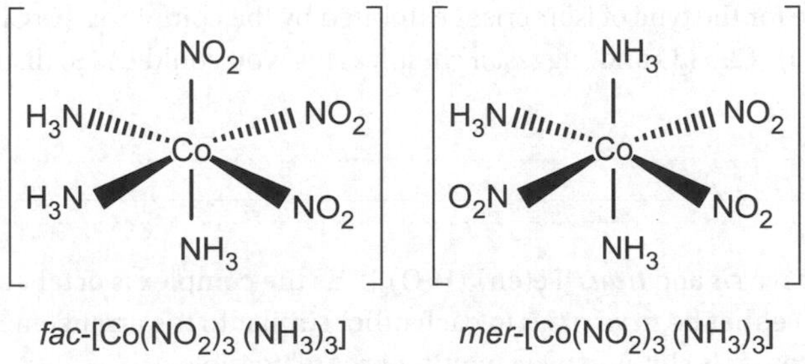

Figure 5.9 *fac* and *mer* isomers of $[Co(NO_2)_3(NH_3)_3]$.

(b) $[CoCl_2(en)_2]^+$ is an octahedral complex with two unidentate and two bidentate ligands. Because there are two ligands of the same type these can either be arranged *cis* to each other or *trans*, and so the complex shows *cis/trans* isomerism as shown in Figure 5.10. The *cis* complex can also exist as optically active enantiomers as we have seen above.

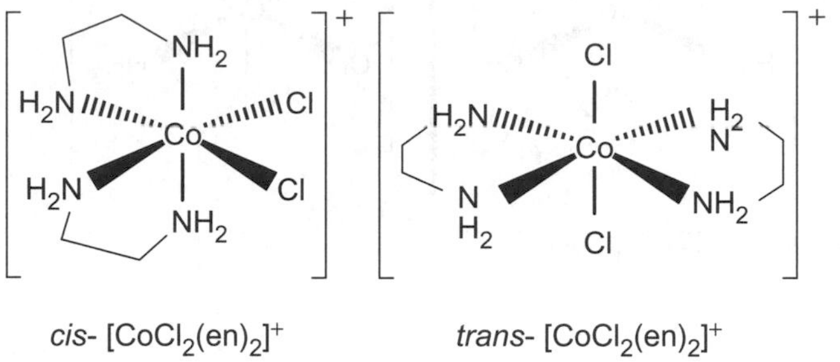

Figure 5.10 *cis*-$[CoCl_2(en)_2]^+$ and *trans*-$[CoCl_2(en)_2]^+$.

(c) $Fe(en)_3$ contains three bidentate ethane-1,2-diamine ligands. This molecule has a chiral centre because it has a non-superimposable mirror image. This is quite difficult to see on paper but much more obvious if you build a model. The two isomers are shown in Figure 5.11.

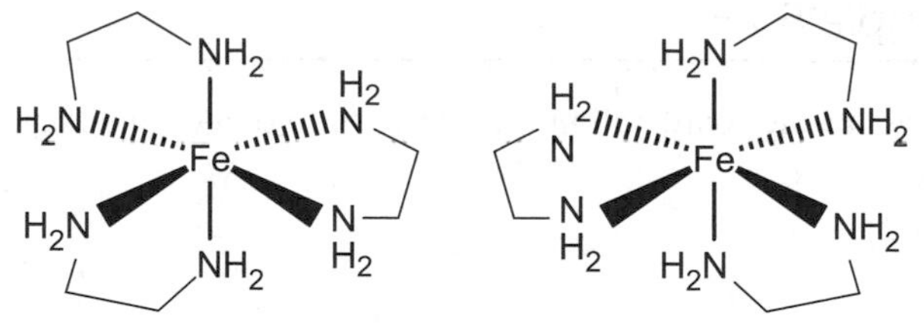

Figure 5.11 Enantiomers of $Fe(en)_3$.

Worked example 5.7C

(a) Give the name for the type of isomerism exhibited by the complex $[Fe(en)_2(H_2O)_2]^{2+}$. State how many isomers are possible and sketch them.

(b) Give the name for the type of isomerism exhibited by the complexes $[CrCl_2(H_2O)_4]Cl.2H_2O$ and $[CrCl(H_2O)_5]Cl_2.H_2O$ and suggest a chemical test you could use to distinguish between them.

Solution

(a) Here we can have *cis* and *trans* $[Fe(en)_2(H_2O)_2]^{2+}$ as the complex is octahedral. The two H_2O molecules can either be *trans* or *cis* to each other similar to the arrangements in Figure 5.10. The *cis* isomer is chiral so two enantiomers are possible.

(b) These complexes have chloride ions both as ligands and counter ions (outside of the coordination sphere of the Cr). The complexes also have water molecules both as ligands and molecules of water of crystallization. These water molecules can exchange with the chloride ligands in the coordination sphere of the Cr ion and so the degree of hydration changes. This means that hydrate isomers are possible.

To distinguish between the isomers, titrate with $AgNO_3$ of known concentration to determine the moles of Cl^- per mole of complex. The silver (Ag^+) will only precipitate the chloride ions that are present outside the coordination sphere of the complex and so a 1:1 molar equivalent of $AgNO_3$ will be required with $[CrCl_2(H_2O)_4]Cl.2H_2O$ and a 2:1 molar equivalent with $[CrCl(H_2O)_5]Cl_2.H_2O$.

Note that the oxidation state of the Cr is unchanged in the complexes—only the overall charge on the complex cation changes.

Question 5.7

State the type of isomerism possible in the following complexes and sketch two possible isomers of each complex.

(a) $[Cr(ox)_2(H_2O)_2]^-$

(b) $[Pt(SCN)_2(PMe_3)_2]$

(c) $[CoF_3(H_2O)_3]$

Question 5.8

Which of the following complexes can exhibit optical isomerism?

(a) $[Mn(en)_2(H_2O)_2]$

(b) $[Cr(acac)_3]$

(c) *trans*-$[CoCl_2(en)_2]Cl$

(d) $[Ni(bipy)_3]^{2+}$

5.8 Crystal-field theory

Crystal-field theory is an electrostatic bonding model that assumes that metal ions in coordination complexes are surrounded by ligands that act as negative point charges. Electrostatic repulsion exists between the electrons of the central metal ion (in *d* orbitals of transition metal ions) and the electrons of the ligands. Crystal-field splitting occurs because the ligand field is not symmetrical with respect to the electrons in the *d* orbitals of the central metal ion and therefore the electrons in these orbitals experience different amounts of repulsion depending upon the location of the ligands with respect to the orbitals.

Crystal-field theory is a model that considers the bonding in coordination complexes to be wholly ionic. An alternative approach that assumes covalent interactions between the metal and the ligands is known as molecular orbital (MO) theory. For transition metal complexes the ligand group orbital (LGO) approach is generally used. This is the approach mentioned in Chapter 2 in which the ligand orbitals are collected into groups with appropriate symmetry to overlap with the valence metal orbitals. The extension of MO theory to octahedral complexes is quite involved and outside the scope of this book.

Octahedral complexes

In an octahedral ligand field the metal *d* orbitals split into two groups, the e_g set and the t_{2g} set. The e_g set contains the d_{z^2} and $d_{x^2-y^2}$ orbitals and the t_{2g} set contains the d_{xy}, d_{yz}, and d_{xz} orbitals. When in an octahedral ligand field the e_g orbitals are destabilized as these are the orbitals that are oriented along the Cartesian axes and directly towards the point charges representing the ligands and so experience an increase in energy. The t_{2g} set of orbitals are oriented between the axes and are more distant from the ligands and so experience a relatively weaker destabilization. The energy difference between the two sets of orbitals is termed the crystal-field splitting energy, Δ_o when in an octahedral field. Figure 5.12 shows the splitting of the *d* orbitals in an octahedral field.

Different ligands split the energies of the *d* orbitals by different amounts producing different values for the splitting energy Δ_o. Ligands that cause a large splitting are called **strong field** ligands and those that cause a small splitting are called **weak field** ligands. The ordering of

The symbol Δ_o or Δ_{oct} stands for 'delta octahedral'.

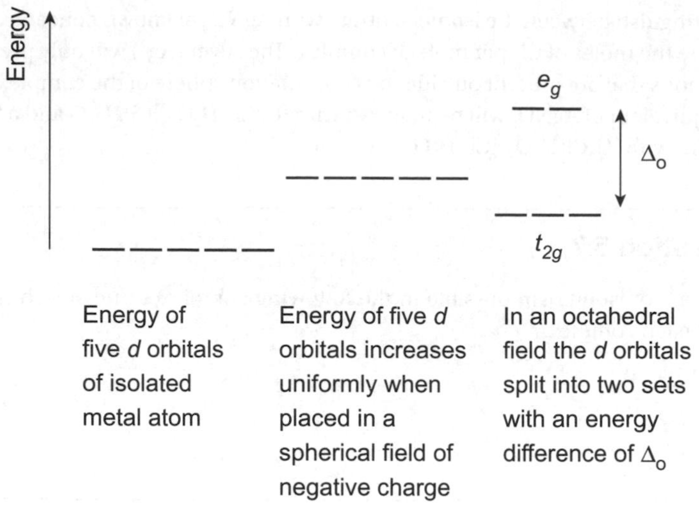
Energy
e_g
Δ_o
t_{2g}
Energy of five d orbitals of isolated metal atom
Energy of five d orbitals increases uniformly when placed in a spherical field of negative charge
In an octahedral field the d orbitals split into two sets with an energy difference of Δ_o

Figure 5.12 Splitting of the *d* orbitals in an octahedral field.

ligands is dependent upon the degree to which they can split the energies of the *d* orbitals and is given by the **spectrochemical series**. For common ligands the spectrochemical series is given below:

Weak field $I^- < Br^- < SCN^- < Cl^- < F^- < OH^- < H_2O < NCS^- < py < NH_3 < en < PR_3 < CN^- < CO$ Strong field

Small Δ Large Δ

In populating the orbitals the valence electrons of the metal fill the lower energy, t_{2g}, orbitals first. When each of the t_{2g} orbitals is half filled the next electron has the possibility of either entering one of the e_g orbitals, in which case we get a **high-spin** complex, or pairing up in a t_{2g} orbital to produce a **low-spin** complex, as in Figure 5.13 for a d^4 metal ion.

High- and low-spin complexes are possible for metals with *d* electron configurations from d^4 to d^7. Whether the complex is high or low spin depends upon a variety of factors including the nature of the ligand, the pairing energy (PE), the metal centre, and its oxidation state. The pairing energy (PE) is the amount of energy required when two electrons pair in the same orbital. Various factors mean that this process requires energy compared to when electrons occupy single orbitals. The pairing energy is the energy required to pair electrons. In some d^n configurations (when $n>5$) a number of electrons must pair up. For *d* block elements there can be a maximum of ten valence electrons and only five orbitals.

When calculating the crystal-field stabilization energy we only need to include the pairing energy if there is another arrangement where the electrons are not paired. For example, in an

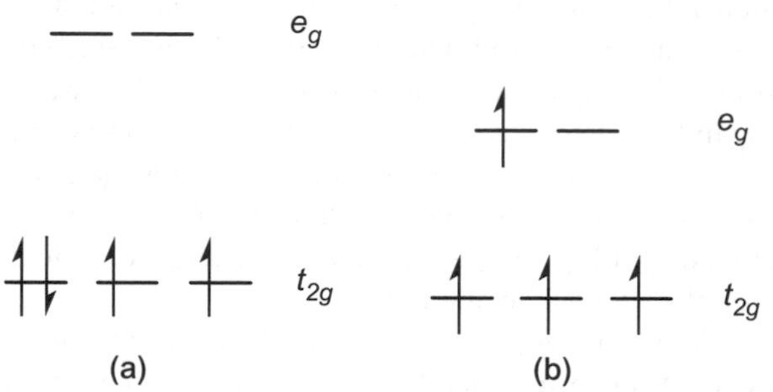
e_g
t_{2g}
(a)
e_g
t_{2g}
(b)

Figure 5.13 (a) A low-spin (strong field) d^4 arrangement and (b) a high-spin (weak field) d^4 arrangement.

octahedral complex with four d electrons we can either arrange the electrons as: $t_{2g}^{3}e_{g}^{1}$ or $t_{2g}^{4}e_{g}^{0}$. In the first arrangement we have no pairs of electrons but in the second arrangement we have one pair. Thus we would need to include one pairing energy, PE, when calculating the difference in energy between the two electron arrangements.

Tetrahedral complexes

A consideration of the orientation of the ligands with respect to the five d orbitals of the metal can be used to derive the energy level splitting diagrams for a variety of different geometries. Tetrahedral complexes have no ligands aligned on any of the Cartesian axes. The energy level splitting diagram for a tetrahedral complex is the inverse of that for an octahedral complex, Δ_t, as shown in Figure 5.14. You should note that the splitting energy for a tetrahedral metal complex is about 4/9 the value for an equivalent octahedral complex with the same ligands in the same oxidation state (if this were possible to synthesize). This is first because there are only four ligands as compared to six in an octahedral complex—so less interaction between the d orbitals and ligands. Secondly none of the ligands are along the Cartesian axes and this again reduces the interaction possible with the d orbitals. Tetrahedral complexes are therefore generally high spin.

Δ_t or Δ_{tet} stands for 'delta tetrahedral'.

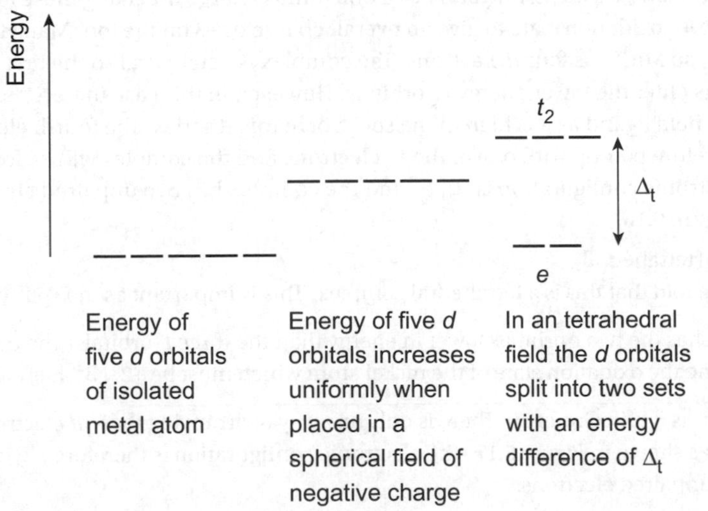
Energy
t_2
Δ_t
e
Energy of five d orbitals of isolated metal atom
Energy of five d orbitals increases uniformly when placed in a spherical field of negative charge
In an tetrahedral field the d orbitals split into two sets with an energy difference of Δ_t

Figure 5.14 Splitting of the d orbitals in a tetrahedral field.

The arrangement of electrons in d orbitals is critical to the spectroscopic and magnetic properties of coordination complexes. Questions involving this topic are varied and exhaustive but they all begin by calculating the number of valence d electrons in the complex. This usually requires a knowledge of the oxidation state of the metal. After this the next step generally involves a consideration of the geometry of the complex and the type of ligands. The next set of worked examples will demonstrate a variety of different types of questions. For a full explanation of the theory you should consult your course text book.

Worked example 5.8A

Determine the electronic configuration in terms of $t_{2g}^{x}e_{g}^{y}$ or $t_{2}^{x}e^{y}$ in each of the following coordination complexes:

(a) $[Fe(H_2O)_6]^{2+}$

(b) $[Fe(H_2O)_6]^{3+}$

(c) $K_3[Mn(CN)_6]$
(d) $[NiCl_4]^{2-}$ (tetrahedral)
(e) $Na_3[CuF_6]$

Solution

(a) $[Fe(H_2O)_6]^{2+}$

Fe is in oxidation state +2. This means it is a $3d^6$ ion (Fe is in Group 8). The water ligand is a **weak field** ligand and so causes a small splitting and a **high-spin** complex. The fourth and fifth electrons will therefore occupy the e_g orbitals and then the final electron pairs up in the lower t_{2g} orbital, as shown in Figure 5.15. The electronic configuration is therefore $t_{2g}^4 e_g^2$ and there are four unpaired electrons.

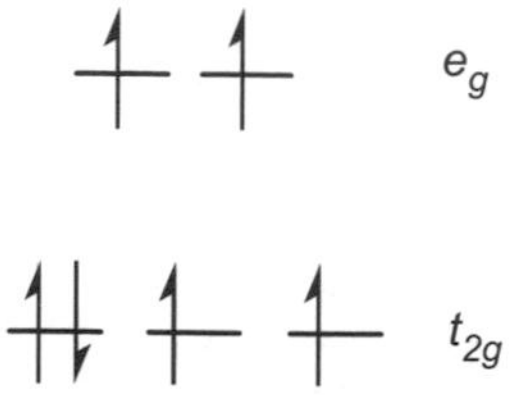

Figure 5.15 Electron arrangement in the 3*d* orbitals of high spin $[Fe(H_2O)_6]^{2+}$.

(b) $[Fe(H_2O)_6]^{3+}$

Here Fe is in the +3 oxidation state so there are five valence electrons, $3d^5$. Again we have a weak field ligand and so the complex is high spin. There is one less electron than in Fe^{2+} and so the electronic configuration is $t_{2g}^3 e_g^2$ with a total of five unpaired electrons.

(c) $K_3[Mn(CN)_6]$

The complex ion here has a −3 charge which is counter-balanced by the three potassium ions. As the CN^- ligand has a one minus charge the manganese ion must be in the +3 oxidation state to give an overall charge of −3 on the ion. Manganese is Group 7, so Mn^{3+} has four *d* electrons. The complex is octahedral so the first three electrons enter the lower energy t_{2g} orbitals. However, in this case the CN^- ligand is a strong field ligand as it is high in the spectrochemical series. The fourth electron will therefore pair up with one of the t_{2g} electrons and the complex will be low spin. The electronic configuration is: $t_{2g}^4 e_g^0$ and the complex has two unpaired electrons as in Figure 5.16.

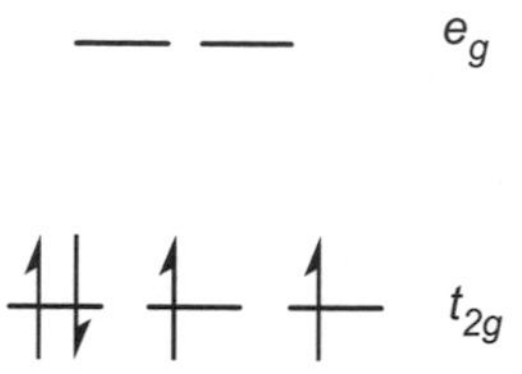

Figure 5.16 Electron arrangement in the 3*d* orbitals of low spin $K_3[Mn(CN)_6]$.

(d) $[NiCl_4]^{2-}$ (tetrahedral)

We are told that this is a tetrahedral complex. This is important as the splitting energy diagram has the two *e* orbitals lower in energy than the three t_2 orbitals. The next step is to determine the oxidation state of the nickel atom which must be +2. Ni^{2+} has eight valence *d* electrons as it is in Group 10. There is only one way of arranging these *d* electrons over five orbitals as shown in Figure 5.17. The electronic configuration is therefore $e^4 t_2^4$ and there are two unpaired electrons.

> The labels for the two sets of orbitals in tetrahedral geometry are t_2 and *e* as they are not 'gerade' as there is no centre of symmetry in a tetrahedral complex.

> Ni is commonly in the +2 oxidation state, apart from with strongly oxidizing ligands such as F^-.

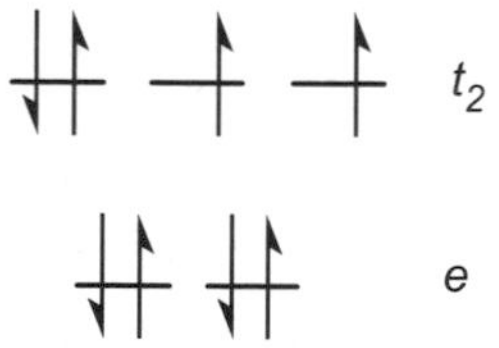

Figure 5.17 Electron arrangement in the 3*d* orbitals in $[NiCl_4]^{2-}$.

(e) $Na_3[CuF_6]$

The complex ion in this salt is $[CuF_6]^{3-}$ which has a copper ion in the +3 oxidation state which is not common. (Cu is normally in the +2 oxidation state but can be oxidized by strongly electronegative elements such as F). Cu^{3+} has eight valence electrons and the ion is octahedral. There is only one way of arranging eight electrons over five orbitals and so the electronic configuration is $t_{2g}^6 e_g^2$.

Question 5.9

For each of the following complexes determine the oxidation state of the metal and the number of *d* electrons. Predict whether the complex will be low or high spin and sketch the energy level splitting diagram.

(a) $[CoF_6]^{3-}$
(b) $[Cr(NH_3)_6]^{3+}$
(c) $Ni(CO)_4$

Determining the crystal-field stabilization energy of a complex

In an octahedral complex when the *d* orbital energy levels split in energy due to the surrounding ligand field the overall energy of the orbitals cannot change, so the amount by which the three t_{2g} orbitals are stabilized must be the same as the amount by which the two e_g orbitals are destabilized. As there are five *d* orbitals in total the t_{2g} orbitals are said to be stabilized by $2/5\Delta_o$ and the e_g orbitals are destabilized by $3/5\Delta_o$, as shown in Figure 5.18.

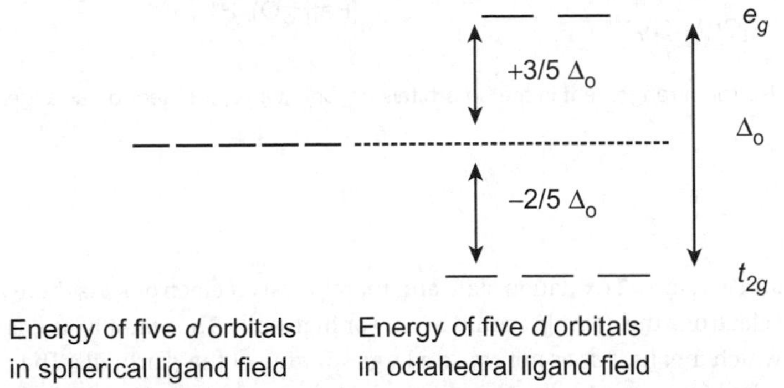

Figure 5.18 The total stabilization of the three t_{2g} orbitals is the same as the destabilization of the two e_g orbitals.

The total stabilization experienced by a complex with a certain number of *d* electrons can therefore be determined by a consideration of the number of *d* electrons and their arrangement within the t_{2g} and e_g orbitals along with the total splitting energy of the orbitals in the ligand field, Δ_o. This stabilization is known as the **crystal-field stabilization energy (CFSE)** and is an important measure of the stability of the ion in a particular complex, and has an impact upon the physical and chemical properties of the complex.

Another energy term that we must introduce here is the pairing energy, PE. If the complex is forced to adopt a low-spin configuration because the splitting energy is large we also have to consider the pairing energy, PE, that must be overcome when the electrons occupy the same orbital. Whether the complex adopts a low-spin or high-spin configuration is therefore dependent upon a balance between the stabilization energy, Δ, and the pairing energy. If Δ is large (strong field) and the pairing energy low the complex will be low spin. If Δ is small (weak field) and the pairing energy high the complex will be high spin.

➔ It is very easy to confuse the terms when discussing crystal-field theory. Remember strong field = low spin and weak field = high spin.

In calculating the resultant CFSE the pairing energy included is a result of the total number of pairs of electrons over and above the minimum number of pairs for that d^n configuration. This will become clear in the examples below.

Worked example 5.8B

Calculate the crystal-field stabilization energy for the following complexes in terms of Δ_o and the pairing energy, PE.

(a) $[Cr(H_2O)_6]^{3+}$

(b) $[Fe(H_2O)_6]^{2+}$

a) $[Co(CN)_6]^{3-}$

Solution

(a) $[Cr(H_2O)_6]^{3+}$

In order to determine the CFSE we must first work out whether the complex is high or low spin and then arrange the electrons in the *d* orbitals. Cr^{3+} is d^3 and so all three electrons are in the t_{2g} orbitals and there are no paired electrons, as shown in Figure 5.19a. The crystal-field stabilization energy is therefore:

$$\text{CFSE} = 3 \times -2/5\Delta_o = -6/5\ \Delta_o$$

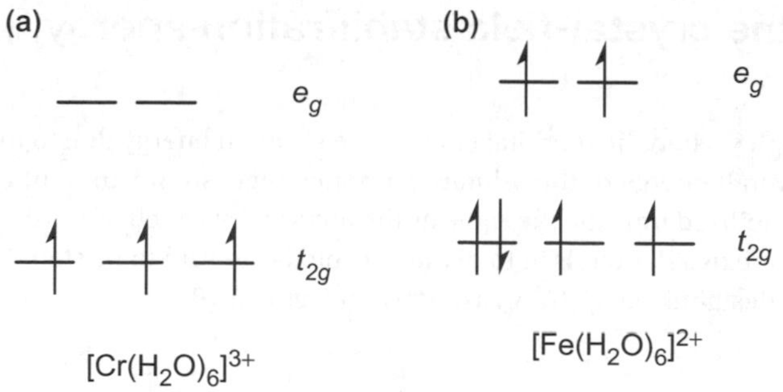

Figure 5.19 (a) Electron arrangement in the 3*d* orbitals of $[Cr(H_2O)_6]^{3+}$. (b) Electron arrangement in $[Fe(H_2O)_6]^{2+}$.

Notice that although there is one pair of electrons in the t_{2g} set there must always be at least one pair of electrons with d^6 configuration and so there is no need to include the pairing energy.

A high-spin d^6 configuration is shown in Figure 5.19b.

(b) $[Fe(H_2O)_6]^{2+}$

This ion has Fe in the +2 oxidation state and therefore six *d* electrons. As there are more than three *d* electrons the complex could be low or high spin. The water ligand is a weak field ligand which means it does not cause a large splitting of the *d* orbitals. The complex will therefore be high spin with the electron arrangement shown in Figure 5.19b.

The CFSE is given by: $\text{CFSE} = 4 \times -2/5\,\Delta_o + 2 \times +3/5\,\Delta_o = -2/5\,\Delta_o$

(c) $[Co(CN)_6]^{3-}$

In this complex Co is in the +3 oxidation state and is therefore d^6. The ligand CN^- is a strong field ligand and so Δ is large. The six electrons will pair up in the lower energy t_{2g} orbitals. The CFSE is given by:

$$\text{CFSE} = 6 \times -2/5\Delta_o + 2\text{PE} = -12/5\,\Delta_o + 2\text{PE}$$

In this case of low-spin d^6 we must include two pairing energy quantities, as the minimum number of pairs of electrons for d^6 is one and so the additional number for low-spin d^6 is two PE.

Question 5.10

Determine the crystal-field stabilization energy for the following complexes in terms of Δ_o or Δ_t and the pairing energy, PE.

(a) $[Cr(en)_3]^{3+}$
(b) $[MnBr_4]^{2-}$
(c) $[Fe(CN)_6]^{3-}$
(d) $[Co(NH_3)_6]^{3+}$

Question 5.11

Determine the number of unpaired electrons in each of the following complex ions:

(a) $[ScF_6]^{3-}$
(b) $[Mn(CN)_6]^{4-}$
(c) $[FeF_6]^{4-}$
(d) $[RuCl_6]^{3-}$

Square planar complexes

There are two common geometries for four-coordinate transition metal complexes: tetrahedral and square planar. We saw earlier how the splitting of the energies of the five *d* orbitals in a tetrahedral ligand field is the inverse of that in an octahedral field. If the complex is square planar the energy levels of the *d* orbitals have a different arrangement.

We can envisage square planar geometry as being derived from octahedral geometry if the axial ligands along the *z* axis are removed. Removal of the ligands on the *z* axis results in a decrease in the interaction of these ligands with the metal *d* orbitals along this axis such that any orbital with a component along the *z* axis is stabilized.

In the e_g set the energy of the d_{z^2} orbital decreases and that of the $d_{x^2-y^2}$ orbital increases. The energies of the t_{2g} orbitals are also affected because they are no longer equivalent in the square planar ligand field. The energy of the d_{xy} orbital is increased whilst that of the d_{xz} and d_{yz} orbitals decreases by half the amount the d_{xy} orbital increases. In a square planar complex with a relatively strong field the energy of the d_{xy} orbital is typically higher than that of the d_{z^2} orbital, as shown in Figure 5.20.

The stabilization energy shown in Figure 5.20 is labelled Δ_{sp} and is relatively large—in fact, usually larger than Δ_o for the same metal ion with the same ligands. Thus a particularly stable electronic configuration is seen in d^8 complexes as the eight electrons fill the four lowest energy orbitals. This is why four-coordinate complexes containing metal ions with eight *d* electrons are frequently square planar, especially with strong field ligands or heavier *d*-block metals. Steric effects can be greater in square planar complexes as the L–M–L angle is smaller (at approximately 90°) rather than 109° as in tetrahedral complexes. However, if the crystal-field stabilization energy is high the complex is likely to be square planar.

A common dilemma is deciding whether a four-coordinate complex is tetrahedral or square planar. Sometimes magnetic properties will help. As we saw in Chapter 1, species with unpaired electrons are paramagnetic and so have a permanent magnetic moment. If all the electrons are paired, however, there will be no magnetic moment. Four-coordinate complexes with eight *d* electrons are often square planar, but not always. So, for example, $[Ni(CN)_4]^{2-}$ is tetrahedral whereas $[Pt(CN)_4]^{2-}$ and $[Pd(CN)_4]^{2-}$ are square planar. The heavier *d*-block metals have more radially expanded 4*d* and 5*d* orbitals which experience

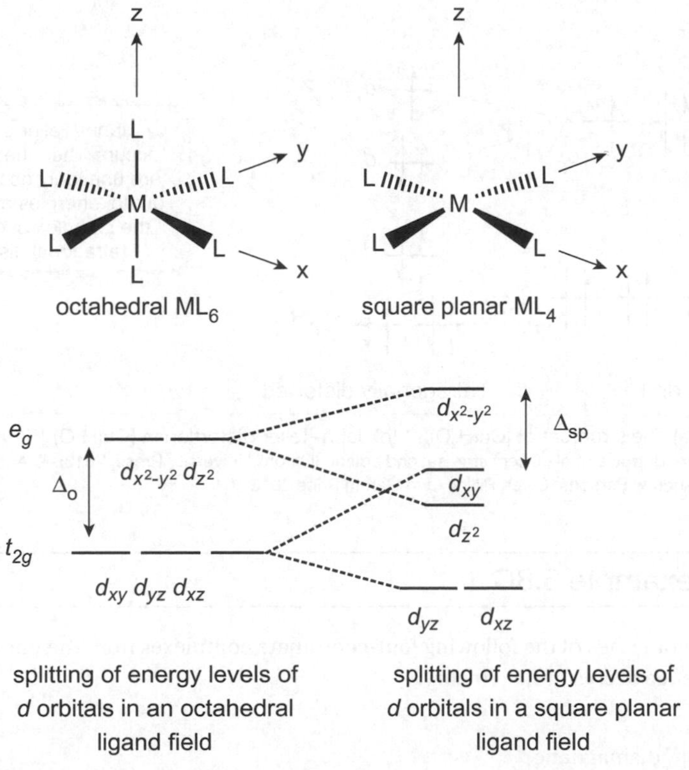

Figure 5.20 Energy level splitting diagram for a square planar ligand field.

a greater interaction with the ligands and so the stabilization energy is greater, thus resulting in square planar complexes with the electrons in lower energy orbitals.

Jahn–Teller distortion

We use the term **degenerate** to indicate that two or more states have the same energy. The term 'doubly degenerate' means that two states (such as electron arrangements in orbitals) have the same energy and the term triply degenerate means three states have the same energy.

In complexes with more than one arrangement of electrons in the ground state, both of which are **degenerate**, **Jahn–Teller distortion** occurs.

This results in the complex distorting so that one of the orbitals has lower energy than the other and the occupation of this orbital leads to further stabilization of the complex. The effects of Jahn–Teller distortion are most important for the electronic configurations d^4 (high spin) and d^9 when there is either one or three electrons in the e_g orbitals respectively. The effect is most pronounced with electrons in the e_g orbitals as these are oriented directly towards the ligands in octahedral complexes.

The complex ion $[Cu(H_2O)_6]^{2+}$ is the classic example of Jahn–Teller distortion (Figure 5.20). Cu^{2+} has nine *d* electrons and so all orbitals are doubly filled apart from one of the e_g orbitals, leaving a hole in either the $d_{x^2-y^2}$ or d_{z^2} orbital. Two degenerate arrangements of electrons are possible, and the complex distorts from regular octahedral geometry to *tetragonal* geometry. In this case $[Cu(H_2O)_6]^{2+}$ has two long axial and four short equatorial Cu-O bonds. As a result the d_{z^2} orbital drops slightly in energy and the $d_{x^2-y^2}$ orbital increases in energy. The d_{z^2} orbital is doubly filled and there is therefore an overall lowering in energy of the complex.

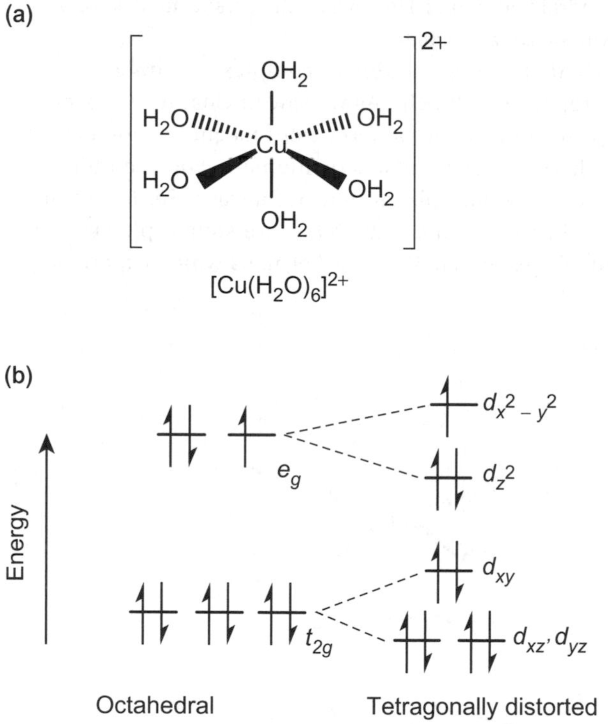

Jahn–Teller distortion occurs when the e_g orbitals are unevenly occupied. The orbital energies are split and the complex undergoes a tetragonal distortion.

Figure 5.21 (a) The structure of $[Cu(H_2O)_6]^{2+}$ (b) Jahn–Teller distortion in $[Cu(H_2O)_6]^{2+}$.

Worked example 5.8C

Predict the geometries of the following four-coordinate complexes from the information given and a knowledge of crystal-field theory:

(a) $[CoCl_4]^{2-}$

(b) $[Ni(CN)_4]^{2-}$ diamagnetic

(c) $[AuCl_4]^-$
(d) $[IrCl(CO)(PPh_3)_2]$

Solution

(a) $[CoCl_4]^{2-}$
This complex ion has Co in the +2 oxidation state and so is d^7. The Cl^- ligand is a weak field ligand and Co a first row transition metal. Therefore the complex is likely to be tetrahedral.
(b) $[Ni(CN)_4]^{2-}$ diamagnetic
Here Ni is in the +2 oxidation state and so is d^8. The complex is diamagnetic. This means that all the electrons are paired. It therefore must be square planar and not tetrahedral.

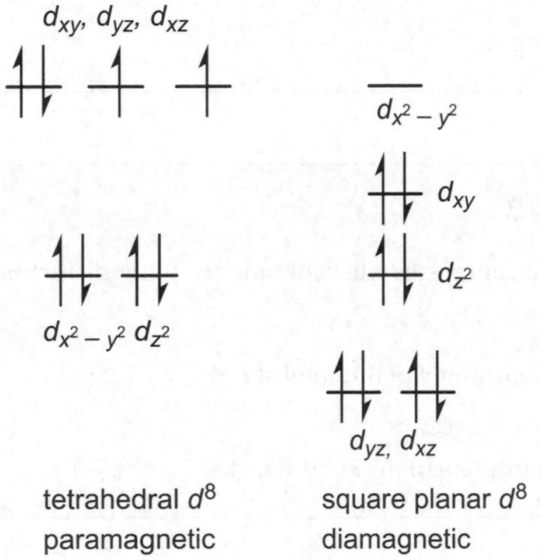

(c) $[AuCl_4]^-$
In this complex Au is in the +3 oxidation state and is therefore d^8. Because this is a third row transition metal complex it is likely to be square planar.
(d) $[IrCl(CO)(PPh_3)_2]$
This complex contains Ir in the +1 oxidation state which is therefore d^8. Again it contains a third row transition element as well as strong field ligands (CO and PPh_3). It is therefore square planar.

Worked example 5.8D

Which of the following complexes would you expect to show Jahn–Teller distortion?
(a) octahedral CrF_2
(b) $CsCuCl_3$ (the $[CuCl_3]^-$ ion has octahedrally coordinated Cu^{2+} ions in the solid state)
(c) $[Cr(H_2O)_6]^{3+}$

Solution

(a) CrF_2 consists of a lattice of octahedrally coordinated Cr^{2+} ions. Cr^{2+} is d^4 and as F^- is a weak field ligand the ion is high spin with one unpaired electron in the e_g orbitals. The uneven occupancy of the e_g orbitals means that tetragonal distortion will occur and so Jahn–Teller distortion is observed. The effect is less pronounced with a d^4 ion than d^9 as there is only one electron in the e_g orbitals with d^4 configuration, whereas there are three with the d^9 configuration.

(b) $CsCuCl_3$ has Cu^{2+} ions which are d^9. As with $[Cu(H_2O)_6]^{2+}$ shown in Figure 5.21 the Cu–Cl octahedral ion will distort to a structure with two elongated axial bonds and four shorter equatorial bonds . The complex is therefore expected to show Jahn–Teller distortion.

(c) $[Cr(H_2O)_6]^{3+}$ has Cr^{3+} ions which are d^3. All three t_{2g} orbitals are therefore evenly filled and there are no degenerate electron arrangements so no Jahn–Teller distortion is possible.

Question 5.12

Predict the geometries of the following four-coordinate complexes from the information given and a knowledge of crystal-field theory:

(a) $[NiBr_4]^{2-}$

(b) $[PtCl_4]^{2-}$

(c) $[Ni(CO)_4]$ diamagnetic

Question 5.13

Which of the following complexes will exhibit Jahn–Teller distortion?

(a) Octahedral MnF_3

(b) K_2CuF_4 (Cu is six coordinate in the solid state)

(c) $[Ni(H_2O)_6]^{2+}$

(d) $KCrF_3$ (Cr is six coordinate in the solid state).

5.9 Colour in coordination complexes of *d*-block metals

A very interesting property of coordination complexes is that they are often coloured. An understanding of crystal-field theory and the spectrochemical series allows us to begin to explain the colours observed.

The colour of a coordination complex arises when visible light is incident on the solid or on a solution of the complex. Electrons in lower energy orbitals are excited by the incident photons and promoted to a higher energy level. In the case of octahedral transition metal coordination complexes the electron is generally promoted from a lower energy t_{2g} orbital to a higher energy e_g orbital. Certain specific wavelengths of the light are absorbed by the complex. This means that the light that is transmitted (passes through) the solution is made up of the remaining wavelengths. A colour wheel can be used to show the colour of the remaining light.

The size of the energy difference between the t_{2g} and e_g orbitals determines the energy and wavelength of the light absorbed. So the nature of the ligands in a complex will affect the colour perceived. A strong field ligand such as CN^- will cause a greater splitting between the t_{2g} and e_g orbitals than a weak field ligand such as H_2O. The larger splitting results in higher energy (smaller wavelength) photons required to excite an electron between the levels.

A further factor that affects the colour of a complex is the oxidation state of the metal ion. For the same metal ion, the higher oxidation state ion will have a greater effective nuclear charge, Z_{eff}, and therefore greater stabilization energy (due to greater interaction between ligand electrons and *d* orbitals). This is associated with higher energy light absorbed and lower energy (longer wavelength) transmitted.

There is a problem with this theory in that it doesn't explain the very intense colours of some transition metal complex ions, such as MnO_4^- which has no *d* electrons [MnO_4^- has Mn(VII) which is d^0]. The colour in this cases arises from the formation of a charge-transfer transition (met in Chapter 3) where the lone pairs of the electronegative O^{2-} ligands are promoted into a low-lying, Mn-centred orbital.

Complexes with charge-transfer transitions often have very intense colours; the MnO_4^- ion is deep purple, I_2 in water is dark brown. The intensity in colour of the solution is related to the **absorbance, *A***, which is expressed by the Beer–Lambert law: $A = \varepsilon c l$.

➔ Complexes having charge-transfer transitions from the ligand to the metal or vice versa are often called 'charge-transfer complexes'.

In this equation c is the concentration of the solution, l is the path length of the cell or container, and ε is a quantity called the molar absorptivity. The value of ε is affected by the selection rules which govern whether an electronic transition is allowed or not.

The spin selection rule states that for a transition to be allowed the spin of the electron should not change, so $\Delta S = 0$ as shown in Figure 5.22.

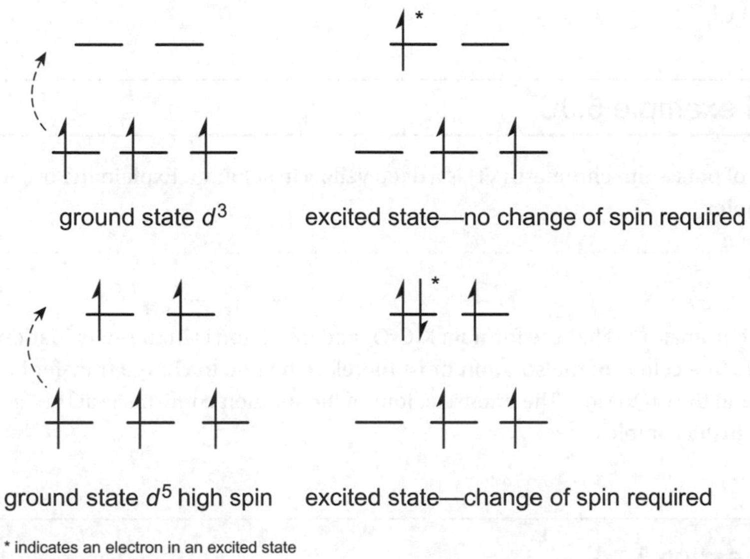

Figure 5.22 The spin selection rule.

The Laporte selection rule states that the quantum number l must change by + or −1 for a transition to be allowed. So an $s \rightarrow p$ transition is allowed but not a $d \rightarrow d$ transition. Another way of expressing this is that $g \rightarrow g$ and $u \rightarrow u$ transitions are forbidden but $g \rightarrow u$ transitions are allowed. This would mean that all electronic transitions between d orbitals in transition metal complexes are forbidden. However, **spin-orbit coupling** and **vibronic coupling** lead to orbital mixing of the d and p orbitals, giving the d orbitals some p character and therefore allowing the transition. This explains why the colours of many transition metal complexes are relatively weak.

Worked example 5.9A

The $[Ti(H_2O)_6]^{3+}$ ion is a pale violet colour in solution. Explain how the colour arises. State whether the transition is allowed or forbidden under the Laporte and spin selection rules.

Solution

The $[Ti(H_2O)_6]^{3+}$ ion has Ti(III) and is d^1. The electron is excited from the t_{2g} to the e_g orbital and certain wavelengths of light are absorbed. The wavelength of the light absorbed is in the yellow/green region of the electromagnetic spectrum so that the light transmitted appears violet.

The electronic transition is between d orbitals and so is Laporte forbidden. However, the spin of the electron remains the same (see Figure 5.22) and so is spin allowed.

➔ In the excited state the electron configuration is $t_{2g}^{\ 0}\ e_g^{\ 1}$. The uneven filling of electrons in the e_g orbitals leads to Jahn–Teller distortion in the excited state and a loss of degeneracy of the e_g orbitals. This results in the absorption spectrum showing a small shoulder on the main peak.

A common point of confusion is explaining the origin of colour due to **emission** and **transmission**. In an **emission** spectrum light energy is being released by electrons returning to lower energy levels and therefore **emitting** energy. When considering the colours of coordination complexes in solution the colour is due to the light transmitted through the solution. As we have seen the wavelength of the transmitted light depends upon the wavelengths (or frequencies) of the light that have been absorbed. If the coordination complex is in the solid state then the colour will be due to the wavelengths reflected from the material after certain wavelengths (or frequencies) have been absorbed. Colours of solids can look different to the colour in solution due to particle size. Diffuse reflectance spectroscopy is used to investigate this effect.

Worked example 5.9B

The complex $[Cr(NH_3)_6]^{3+}$ is a yellow-orange colour whereas the complex $[Cr(NH_3)_5Cl]^{2+}$ is purple. Explain the difference in colour of the two complexes.

Solution

Both Cr complexes contain Cr(III) and so the difference in colour must be due to the different ligands. The Cl^- ligand is a weaker field ligand than NH_3 and so causes a smaller splitting of the *d* orbitals in $[Cr(NH_3)_5Cl]^{2+}$. This shifts the energy of the light absorbed to lower frequencies (longer wavelengths). The energy of the light transmitted therefore shifts to higher frequencies (shorter wavelengths) and so the colour changes from yellow-orange in $[Cr(NH_3)_6]^{3+}$ to purple in $[Cr(NH_3)_5Cl]^{2+}$.

Worked example 5.9C

The colour of potassium chromate (VI) is a deep yellow in solution. Explain the origin of colour in this complex.

Solution

Potassium chromate (VI) has the formula K_2CrO_4 and the Cr ion is in the +6 oxidation state and therefore d^0. The colour of the solution must therefore be due to charge transfer between the O^{2-} ligand and the Cr(VI) ion. The intense colour of the solution confirms that this is likely to be a charge-transfer complex.

Question 5.14

A solution of $[V(H_2O)_6]^{2+}$ is violet and a solution of $[V(H_2O)_6]^{3+}$ is green. Explain the difference in colour of the two octahedral vanadium ions and state whether the electronic transition obeys the selection rules.

Question 5.15

Use the colours of the following low-spin Co(III) complexes to suggest the order of the ligands H_2O, Cl^-, and NO_2^- in the spectrochemical series:

$[Co(NH_3)_5(H_2O)]^{3+}$ bright red; $[Co(NH_3)_5Cl]^{2+}$ purple; $[Co(NH_3)_5(NO_2)]^{2+}$ orange.

Question 5.16

The unstable niobium pentahalide $NbCl_5$ is bright yellow in the solid state whereas $TaCl_5$ is white. Explain the origin of, and the difference in colour between, the two complexes.

Turn to the Synoptic questions section on page 148 to attempt questions that encourage you to draw on concepts and problem-solving strategies from several topics within a given chapter to come to a final answer.

Final answers to numerical questions appear at the end of the book, and fully worked solutions appear on the book's website. Go to http://www.oxfordtextbooks.co.uk/orc/chemworkbooks/.

References

IUPAC (1997) *Compendium of Chemical Terminology*, 2nd edn (the 'Gold Book'), compiled by A. D. McNaught and A. Wilkinson (Blackwell Scientific Publications, Oxford). [XML online corrected version: http://goldbook.iupac.org (2006–) created by M. Nic, J. Jirat, B. Kosata; updates compiled by A. Jenkins. ISBN 0-9678550-9-8. doi:10.1351/goldbook.]

Synoptic questions

CHAPTER 1

Question S1.1

(a) Explain why the 2*s* orbital is at a lower energy than the 2*p* orbital for Li, whereas the 2*s* and 2*p* orbitals have the same energy in Li^{2+}.

(b) Give the electronic configurations for the Na and Mg atoms.

(c) Use these configurations to explain why Na^{2+} compounds do not exist and Mg^{+} compounds are unlikely to exist.

(d) Discuss your answer to part (c) in terms of Born–Haber cycles for the proposed compounds $NaCl_2$ and MgCl. You may assume that the ionic radius of Na^{2+} is the same as that of Mg^{2+} and that the lattice enthalpy of MgCl is the same as that of NaCl and that the lattice enthalpy of $NaCl_2$ would be the same as that of $MgCl_2$.

Question S1.2

(a) The Cl–Cl bond in Cl_2 has a bond enthalpy of 242 kJ mol^{-1}. What frequency of photons does this energy value correspond to?
(b) What wavelength of light does this correspond to?
(c) Which part of the electromagnetic spectrum does this energy value fall in?
(d) Draw a molecular orbital energy level diagram for the Cl_2 molecule.
(e) Identify the energy gap corresponding to the HOMO $\rightarrow$ LUMO transition in this molecule.
(f) Would promotion of one electron from the HOMO to the LUMO in Cl_2 be likely to lead to dissociation of the molecule?

CHAPTER 2

Question S2.1

(a) How many electrons are found in the degenerate π^* orbital in O_2?
(b) Sketch the shape of the π^* orbital and show how this orbital might overlap with a *d* orbital on a transition metal atom.
(c) If O_2 coordinates to a transition metal atom, electrons can be transferred from a metal *d* orbital to the π^* orbital on O_2. Explain whether you would expect this process to strengthen or weaken the O_2 bond.
(d) Use the following data to explain why oxygen exists as O_2 molecules rather than single (σ)-bonded O_8 rings analogous to the S_8 rings formed by sulfur: O=O, +498 kJ mol^{-1}; O–O, +144 kJ mol^{-1}.
(e) Suggest why S=S π bonds will be much less strong than O=O π bonds.

Question S2.2

(Also draws on Chapter 5)
(a) Use molecular orbital theory to explain why the colour of the halogens changes from pale yellow (F_2), to green (Cl_2), to reddy-brown (Br_2), to purple (I_2).
(b) Explain why Cu(II) complexes are coloured but Cu(I) complexes are generally colourless.
(c) Explain why $[MnO_4]^-$ has a deep purple colour while $[ReO_4]^-$ is colourless.
(d) Explain why AgI is yellow while AgCl is white; likewise explain why SnI_4 is orange while $SnCl_4$ is colourless.
(e) Explain why $[Mn(H_2O)_6]^{2+}$ has a much less intense colour than does $[MnCl_4]^-$.

Question S2.3

(Also draws on Chapter 5)

State whether or not the following molecules or ions are diamagnetic or paramagnetic and suggest explanations for your answers:

(a) $[Cu(H_2O)_6]^{2+}$
(b) NO
(c) $[NiCl_4]^{2-}$
(d) $[Ni(CN)_4]^{2-}$
(e) $[Pt(NH_3)_2Cl_2]$
(f) O_2
(g) B_2
(h) C_2

CHAPTER 3

Question S3.1

(Also draws on Chapter 4)

(a) Sketch the structure for the unit cell of CsCl.
(b) Suggest an explanation for the fact that TlCl adopts the same structure as CsCl.
(c) Sketch the structure for the unit cell of LiCl.
(d) Sketch the structure adopted by solid $BeCl_2$.
(e) Suggest an explanation for the fact that while CsCl and TlCl have very similar structures, LiCl and $BeCl_2$ have very different structures.

Question S3.2

(a) Why is the 'inert pair effect' much more important in lead chemistry than it is in silicon chemistry?
(b) Explain the observation that while $PbCl_2$ is ionic in the solid state with a structure where each Pb^{2+} ion is coordinated by nine Cl^- ions; $PbCl_4$ exists as a covalent monomer under standard conditions.
(c) Explain the observation that monomeric $PbCl_4$ has a tetrahedral structure in the vapour phase.
(d) Heating solid $PbCl_2$ gives $PbCl_2$ molecules in the vapour phase. Suggest an explanation for the observation that these molecules have a non-linear structure.

CHAPTER 4

Question S4.1

(a) Account for the following trends in first ionization energy:
  i. The increase from oxygen to fluorine.
  ii. The decrease from sodium to potassium.
  iii. The decrease from phosphorus to sulfur.
(b) Use a Born–Haber cycle to discuss why sodium normally exists as Na^+ in most sodium compounds and Na^- compounds are very rare, even though the first ionization potential of sodium:

  $$Na(g) \rightarrow Na^+(g) + e^-$$

  is endothermic (+496 kJ mol^{-1}), while the first electron affinity:

  $$Na(g) + e^- \rightarrow Na^-(g)$$

  is exothermic (–53 kJ mol^{-1}).
(c) Use a Born–Haber cycle to discuss why oxygen often exists as the O^{2-} anion in oxides although the sum of the first two electron affinities of oxygen:

  $$O(g) + 2e^- \rightarrow O^{2-}(g)$$

  is endothermic (+702 kJ mol^{-1}).

This value is the **sum** of the first and second electron affinities of O. The first electron affinity is **exothermic** (–142 kJ mol^{-1}) while the second electron affinity is strongly **endothermic** (+844 kJ mol^{-1}). For the second electron affinity you are adding an electron to an already negatively charged anion.

Question S4.2

(Also draws on Chapter 5)

(a) Write a chemical equation for the decomposition of a group 2 sulfate to an oxide.

(b) Use the Kapustinskii equation to estimate values of the lattice enthalpies of:

i. $CaSO_4$

ii. CaO

iii. $BaSO_4$

iv. BaO

(c) Hence show which of $CaSO_4$ and $BaSO_4$ is more easily decomposed thermally to its oxide.

CHAPTER 5

Question S5.1

Dimethylsulfoxide ($Me_2S{=}O$) is a ligand that can act as an S- or O-donor ligand (known as an ambidentate ligand).

(a) Suggest a structure for $Me_2S{=}O$ using VSEPR theory. Show that there is a lone pair on the S atom.

(b) Explain what is meant by the **hard** and **soft** acids and bases theory as applied to transition metal coordination complexes.

(c) Use this theory to predict which of either the S or an O atom in $Me_2S{=}O$ would coordinate to i. Cd^{2+} and ii. Fe^{3+}.

(d) Use VSEPR theory to predict the structure of the thiocyanate anion $[SCN]^-$.

(e) Use hard and soft acids and bases theory to predict whether it is the S or the N atom that is likely to coordinate to the Fe^{3+} ion in the blood-red complex $[Fe(NCS)(H_2O)_5]^{2+}$ formed when $[NCS]^-$ ions are added to Fe^{3+} ions in aqueous solution.

Question S5.2

(a) An aqueous solution of chromium(III) contains the $[Cr(H_2O)_6]^{3+}$ ion. The Δ_o value for this ion is 17 400 cm^{-1} and the solution is green in colour.

i. Draw the *d* orbital energy level splitting diagram for $[Cr(H_2O)_6]^{3+}$ and insert the appropriate number of *d* electrons into the energy levels.

ii. If 0.880 M ammonia solution is added to the aqueous solution in i. above, the Δ_o value shifts to 21 600 cm^{-1}. Explain this shift in absorption and predict the formula for the complex formed. What colour do you expect the new solution will be?

(b) A further solution contains the $[CrO_4]^{2-}$ ion and is bright yellow in colour. Explain the origin of the colour in this solution.

(c) Dilute sulfuric acid is added to the yellow solution in part (a) above causing the colour to change to a bright orange. Write a balanced chemical equation for this reaction.

(d) Addition of zinc powder to the solution in part (c) causes the colour to change to green. Suggest a possible formula for the species responsible for the green colour and write a balanced chemical equation for the reaction.

Answers

Final answers to questions posed in the text (where they can be given) are presented here. You can find fully worked solutions with many more explanations for every question featured in the book in the *Workbooks in Chemistry* Online Resource Centre. Go to www.oxfordtextbooks.co.uk/orc/chemworkbooks/.

CHAPTER 1

Question 1.1
242 nm

Question 1.2
Ideal speed of electrons = 2.58×10^6 m s^{-1}

The neutron is a much heavier particle and hence requires a lower speed: 1400 m s^{-1}

Question 1.3
436 kJ mol^{-1}

Question 1.4
(a) 515 nm, visible light
(b) 0.11 %

Question 1.5
(a)

n_i	Frequency/Hz
4	1.60×10^{14}
5	2.34×10^{14}
6	2.74×10^{14}

(b) Higher

Question 1.6
(a)
i. Lyman
ii. Balmer
iii. Neither (actually belongs to Paschen series)
iv. Balmer
v. Lyman

Question 1.7
(a)

Orbital	n	l	m_l
2s	2	0	0
2p	2	1	−1, 0, +1
3s	3	0	0
3p	3	1	−1, 0, +1
3d	3	2	−2, −1, 0, +1, +2

(b)

Orbital	Total nodes	Angular nodes	Radial nodes
2s	1	0	1
2p	1	1	0
3s	2	0	2
3p	2	1	1
3d	2	2	0

As expected, the number of nodes increases with increasing principal quantum number

(c) Hydrogen: orbitals with same n are degenerate; Order = (2*s*, 2*p*) < (3*s*, 3*p*, 3*d*).

Question 1.8
(a) 1*s*
(b) 3*p*
(c) 5*d*
(d) 3*d*
(e) 5*f*

Question 1.9
(a) Na: $1s^2 2s^2 2p^6 3s^1$, 1 unpaired e^-
(b) Na^+: $1s^2 2s^2 2p^6$, 0 unpaired e^-
(c) Cl: $1s^2 2s^2 2p^6 3s^2 3p^5$, 1 unpaired e^-
(d) Cl^-: $1s^2 2s^2 2p^6 3s^2 3p^6$, 0 unpaired e^-
(e) Al: $1s^2 2s^2 2p^6 3s^2 3p^1$, 1 unpaired e^-
(f) N^{3-}: $1s^2 2s^2 2p^6$, 0 unpaired e^-
(g) Mg: $1s^2 2s^2 2p^6 3s^2$, 0 unpaired e^-
(h) Mg^{2+}: $1s^2 2s^2 2p^6$, 0 unpaired e^-

Question 1.10
(a) Li: [He] $2s^1$
(b) S^{2-}: [Ar]
(c) Ca: [Ar] $3s^2$
(d) I^-: [Xe]
(e) Si^{2+}: [Ne] $3s^2$
(f) K: [Ar] $4s^1$
(g) Ba^{2+}: [Xe]
(h) H^-: [He]

CHAPTER 2

Question 2.1
(a) Diamagnetic
(b) Paramagnetic
(c) Diamagnetic

Question 2.2
(a) Diamagnetic
(b) 1 (4 bonding pairs of electrons–3 antibonding pairs of electrons)
(c) Stronger—electron removed from antibonding orbital, bond order changes to 1.5
(d) Weaker—electron added to antibonding orbital, bond order changes to 0.5

Question 2.3
We expect to see more examples.

Question 2.4
The 2*s* and 2*p* orbitals are closer in energy in N than in F. Yes, σ–π crossover is more likely in the MO energy level diagram for Cl_2 than for F_2.

Question 2.5
(a) Weaker bonding expected due to poorer overlap between H(1*s*) and Na(3*s*) than Li(2*s*).
(b) i. True
ii. True
iii. False
iv. False
v. True

Question 2.6
(b) Bond order of NO is 2.5
i. Ionization removes an electron from an antibonding orbital and hence NO^+ is expected to have a stronger (bond order = 3) and hence shorter bond.
ii. Addition of an electron to form NO^- places an additional electron into an antibonding

orbital and hence the bond order reduces to 2. This would weaken the bond and hence we would expect it to be longer.

Question 2.7

(a) Orbitals may overlap in phase (bonding), out of phase (antibonding).
(b) The diagram should show bonding, non-bonding, and antibonding molecular orbitals. Only the two lowest energy orbitals are filled.
(c) No significant change.

Question 2.8

(a) Octahedral structure with central S atom.
(b) Worked example 2.9A shows how three $2p_z$ orbitals can overlap to give bonding, non-bonding, and antibonding orbitals. By considering each 'SF_2' unit separately, we treat all the interactions to be between the $2p_z$ orbitals as we are only considering one interatomic axis in each case.
(c) Bonding and non-bonding orbitals should be filled.

Question 2.9

(a) No
(b) Yes (only 1 electron to bond 2 atoms)
(c) No
(d) No
(e) No
(f) Yes (you need to consider the MO diagram: this has two bonding electrons and one antibonding electron so overall there is a bond formed by only one bonding electron, so it is electron deficient)
(g) Yes
(h) Yes. Consideration of the MO diagram shows that there is a net total of one bonding electron in this anion so the ion has an electron deficient bond.

Question 2.10

(a) B_5H_9 has a square-based pyramidal structure and $[B_6H_6]^{2-}$ has an octahedral structure.
(b) B_5H_9: 24 valence electrons present. The 17 covalent bonds would require 34 electrons for standard 2c2e bonds and hence the structure is electron deficient.
(c) $[B_6H_6]^{2-}$: 26 valence electrons present. The 18 covalent bonds would require 36 electrons for standard 2c2e bonds and hence the structure is electron deficient.

Question 2.11

(a) The allyl anion has a bent structure. Lewis (resonance) structures show a double bond between two of the carbon atoms with the anion centred on the other CH_2 group. Alternatively, and more correctly, the minus charge may be considered as being delocalized across all three C atoms.
(b) The $2p_x$ orbitals overlap in a manner analogous to the corresponding orbitals in CO_2 although in the case of the allyl anion, the ion is non-linear.
(c) Counting electrons shows that there is one valence electron on each carbon atom for π bonding. These three electrons fill the bonding molecular orbital and half-fill the non-bonding. Thus adding or subtracting electrons either adds or subtracts an electron from the non-bonding molecular orbital so we do not expect a significant change in bond lengths.

Question 2.12

(a) NH_3 is pyramidal and BCl_3 is planar.
(b) NH_3: sp^3, BCl_3: sp^2

Question 2.13

(a) i. sp^3
ii. sp^2
iii. sp
iv. sp^3
(b) i. sp^3
ii. sp^2
iii. sp^3
iv. sp

Question 2.14

(a) Hybrid approach results in four degenerate bonding orbitals, LGO approach results in one lower energy bonding orbital, and three degenerate higher energy bonding orbitals.
(b) LGO approach.

Question 2.15

(a) Bent
(b) Trigonal pyramidal
(c) Octahedral
(d) See-saw
(e) Trigonal planar
(f) Bent

Question 2.16

(a) Tetrahedral
(b) Octahedral
(c) Linear
(d) Trigonal bipyramidal
(e) Trigonal pyramidal
(f) Tetrahedral
(g) Tetrahedral around each Si atom with two bridging O atoms

CHAPTER 3

Question 3.1

Overall trend is an increase in ionization enthalpy. Anomalies are seen at Mg → Al (Al has a single electron in $3p$ orbital) and at $P \rightarrow S$ ($3p^4$ configuration for S with paired electrons in one p orbital).

Question 3.2

(a) i. Strontium (group 2 versus group 1)
ii. Thallium (group 13 versus group 2)
iii. Strontium (higher in group 2).
(b) i. Moving from group 17 to group 18
ii. Descending group 15
iii. Moving from an element with a $(\ldots)4p^3$ configuration to one with a $(\ldots)4p^4$ configuration

Question 3.3

Due to the balance between the decreasing enthalpies of atomization and ionization down the group, and the decreasing enthalpy of hydration down the group.

Question 3.4

In aqueous solutions, H_2O is preferentially reduced over M^+. For all alkali metals, $E^{\ominus}(H_2O) > E^{\ominus}(M^+)$.

Question 3.5

(a) $MSO_4(s) \xrightarrow{\Delta} MO(s) + SO_3(g)$
(b) $M(OH)_2(s) \xrightarrow{\Delta} MO(s) + H_2O(g)$

As the cation becomes bigger, the sulfate or hydroxide becomes more stable with respect to the oxide.

Question 3.6

The group 2 ions with a 2+ charge have a high charge-to-size ratio and are therefore highly polarizing. This stabilizes the nitride (see answer to Worked example 3.1A) but facilitates breaking of the O–O bond within the peroxide ion.

Question 3.7
(a) Yes; 1:1 stoichiometric adduct
(b) Yes; 1:1 and 1:2 stoichiometric adducts possible

Question 3.8
Both have bridged dimeric structures. Ga_2H_6 is electron deficient with three-centre, two-electron bonds for the bridging hydrogens, while Ga_2Cl_6 has dative bonding and is electron precise.

Question 3.9
$6s^2$ electrons held closely to the nucleus, so not readily oxidized or shared. At the bottom of the group the ns^2 electrons which penetrate to the nucleus feel a stronger nuclear charge. So the effect is more pronounced for Pb than for C. As such Pb^{4+} is relatively unstable and is easily reduced to Pb^{2+}.

Question 3.10
(a) $SnCl_4$: tetrahedral molecules, liquid; $SnCl_2$: polymeric solid of non-linear $SnCl_2$ units, linked by dative covalent bonds.
(b) i. $3SiH_3Cl + NH_3 \rightarrow N(SiH_3)_3 + 3HCl$
ii. $2SiH_3Cl + H_2O \rightarrow (H_3Si)_2O + 2HCl$
iii. $2SiH_3Cl + 2Na \rightarrow Si_2H_6 + 2NaCl$
$N(SiH_3)_3$ has a planar structure, $O(SiH_3)_2$ is non-linear but with a wide bond angle. Si_2H_6 has an electron-precise structure akin to ethane.

Question 3.11
(a) AX_3E_1 configuration. Four pairs of electrons, so shape is based on tetrahedron to maximize distance between pairs of electrons. Lone pair in sp^3 hybrid orbital.
(b) H-bonding
(c) i. pyramidal
ii. higher
(d) i. NH_3
ii. NH_3

Question 3.12
(a) Based on P_4 tetrahedron with edge bridging O atoms P_4O_6 and also terminal O atoms P_4O_{10}
(b) Based on P_4 tetrahedron
(c) Nitrogen—no; Arsenic—possibly
(d) Tetrahedral; each has one P=O group, P–O–H replaced by P–H as sequence followed.
(e) +1, +3, and +5 respectively
(f) Fewer O–H bonds where H^+ ions may be lost as sequence followed.

Question 3.13
(a) O=O
(b) Bent (AX_2E_1)
(c) The free acid cannot be isolated but its salts are well known.
(d) The bond is longer as electrons are added to antibonding orbitals.
(e) Yes

Question 3.14
(a) $2H_2O_2(l) \rightarrow 2H_2O(l) + O_2(g)$
(b) $2Fe^{2+}(aq) + H_2O_2 + 2H^+(aq) \rightarrow 2Fe^{3+}(aq) + 2H_2O(l)$
(c) $2MnO_4^-(aq) + 3H_2O_2(l) \rightarrow 2MnO_2(s) + 2OH^-(aq) + 2H_2O(l) + 3O_2(g)$
(d) $PhSMe(l) + H_2O_2(l) \rightarrow PhS(O)Me(s) + H_2O(l)$
(e) $OCl^-(aq) + H_2O_2(l) \rightarrow Cl^-(aq) + H_2O(l) + O_2(g)$

Question 3.15
(a) F^- is very hard to oxidize. Br^- and I^- can be oxidized by halogens higher up the group, e.g. Cl_2 will oxidize both and Br_2 will oxidize I^-.
(b) The coordinating solvent can donate electrons into the σ* molecular orbital thus forming a charge transfer complex.
(c) The electronegative non-metals show a small electronegativity difference with the halogen atoms and hence lead to covalent compounds. High oxidation state metals are polarizing and have high ionization enthalpies. These factors favour covalent bonding.
(d) The negative charge may be delocalized across more electronegative oxygen atoms stabilizing the anion.

Question 3.16
(a) 24.8% Cu
(b) Hydrated copper(II) sulfate $= CuSO_4.5H_2O = 25.45$ % Cu, therefore 97.4 % pure. Assume no error in measurement and no copper in impurities.

Question 3.17
(a) $Cs^+[XeF_7]^-$
(b) $[XeF_5]^+[SbF_6]^-$
(c) $Kr + HF + O_2$
(d) $IrF_5 + Xe$
(e) $SF_6 + Xe$

CHAPTER 4

Question 4.1
68%

Question 4.2
4.252×10^{-22} g

Question 4.3
(b) 2
(c) 7.85 g cm^{-3}

Question 4.4
(b) 4

Question 4.5
4

Question 4.6
(a) Yes
(b) No
(c) No
(d) No
(e) Yes

Question 4.7
Lattice enthalpy is dependent on charge-to-size ratio.
(a) Cl^- has a higher charge-to-size ratio than Br^-.
(b) Mg^{2+} and O^{2-} have higher charge-to-size ratios than Na^+ and Cl^-.
(c) Enthalpy of formation depends on lattice enthalpy and the enthalpy change associated with formation of the gaseous ions. The magnitude of both quantities decreases down a group, so there is some 'cancelling out'.

Question 4.8
−2260 kJ mol^{-1}

Question 4.9
275 pm

Question 4.10
(b) −283 kJ mol^{-1}
(c) −3982 kJ mol^{-1}

Question 4.11
Born–Landé = −734 kJ mol^{-1}
Kapustinskii = −729 kJ mol^{-1}

Question 4.12
Deviation from ionic bonding model used in Kapustinskii equation increases down the group as ions become more polarizable and more covalency seen in the bonding.

Question 4.13
+266 kJ mol^{-1}. Endothermic, therefore unstable.

Question 4.14
−632 kJ mol^{-1}

CHAPTER 5

Question 5.1
(a) $[Ar]4s^2 3d^8$ and $[Ar]3d^8$
(b) $[Ar]4s^2 3d^5$ and $[Ar]3d^0$
(c) $[Xe]4f^{14} 6s^2 5d^4$ and $[Xe]4f^{14} 5d^3$

Question 5.2
(a) +2 and $[Ar]3d^5$
(b) +2 and $[Ar]3d^8$
(c) +3 and $[Ar]3d^3$
(d) 0 and $[Ar]3d^6$

Question 5.3
(a) Six and octahedral
(b) Four and tetrahedral
(c) Six and octahedral
(d) Six and octahedral

Question 5.4
(a) Pentaamminethiocyanatochromium(III) ion
(b) Tetraamminedinitrito-*N*-cobalt(III) ion or tetraamminedinitrito-κN-cobalt(III) ion
(c) Diaquadioxalatomanganate(II) ion
(d) Tricarbonylbis(triphenylphosphine)iron(0)

Question 5.5
(a) $[CrBr_2(H_2O)_2(NH_2CH_3)_2]NO_3$
(b) $[RuCl(NH_2CH_2COO)(CO)_3]$
(c) $[CoCl_2(en)_2]Cl$

Question 5.6
(a) $[Co(bipy)_3][Fe(CN)_6]$ or $Fe(CN)_4(bipy)]$ $[Co(CN)_2(bipy)_2]$ or $[Fe(bipy)_3][Co(CN)_6]$
(b) $[Pt(NO_2)(NH_3)_3]Br$
(c) $[CoCl(ONO)(NH_3)_4]Cl$
(d) Through the S atom

Question 5.7
(a) *cis/trans*
(b) *cis/trans*
(c) *fac* and *mer*

Question 5.8
(a) Yes
(b) Yes
(c) No
(d) Yes

Question 5.9
(a) +3, 6 *d* electrons, high spin
(b) +3, 3 *d* electrons, neither high nor low spin
(c) 0, 10 *d* electrons, neither high nor low spin

Question 5.10
(a) $-6/5\Delta_o$
(b) Zero
(c) $-2\Delta_o + 2PE$
(d) $-12/5\Delta_o + 2PE$

Question 5.11
(a) Zero
(b) 1
(c) 4
(d) Zero

Question 5.12
(a) Tetrahedral
(b) Square planar
(c) Tetrahedral

Question 5.13
(a) Yes
(b) Yes
(c) No
(d) Yes

Question 5.14
Colour difference due to difference in Δ_0 values causing different wavelengths of light to be absorbed. $[V(H_2O)_6]^{2+}$ and $[V(H_2O)_6]^{3+}$ absorptions both Laporte forbidden but spin allowed.

Question 5.15
$Cl^- < H_2O < NO_2^-$

Question 5.16
Colour due to charge complex formation.
$NbCl_5$ is more readily reduced than $TaCl_5$ and has lower energy transition giving yellow complex. Higher energy transition in uv in Ta complex produces white colour.

Synoptic questions

CHAPTER 1

Question S1.1

(a) Shielding of outermost electron in Li.
(b) Na: [Ne]$3s^1$; Mg: [Ne]$3s^2$
(c) Na^{2+} would require removal of an inner shell electron; Mg^+ has a half-filled 3*s* orbital.
(d) The large second ionization energy of Na (4562 kJ mol^{-1}) cannot be compensated for by the likely lattice enthalpy of the ionic solid $NaCl_2$. A Born–Haber calculation for MgCl would suggest that the enthalpy of formation of this ionic solid is exothermic; however, its enthalpy of formation is much lower than that of $MgCl_2$.

Question S1.2

(a) 6.07×10^{14} Hz
(b) 4.94×10^{-5} m
(c) Visible
(d) See earlier examples
(e) This is from $\pi^*(2p_{xy}) \rightarrow \sigma^*(2p_z)$
(f) No—there is still the same number of antibonding electrons.

CHAPTER 2

Question S2.1

(a) 2
(b) Weaken as going into an antibonding orbital.
(c) Four O_2 molecules have four O=O bonds = 4 × 498 = 1992 kJ mol^{-1}; S_8 has eight O–O bonds = 8 × 144 = 1152 kJ mol^{-1} so less stable overall.
(d) Poorer overlap of *p* orbitals on larger S atoms.

Question S2.2

The colour will be caused by promotion of an electron from the HOMO (π^*) to LUMO (σ^*). The energy gap between HOMO and LUMO decreases as the group is descended. Thus longer wavelength light is absorbed and hence **shorter** wavelength light is **transmitted**.
Cu^{2+} has a d^9 electronic configuration; Cu^+ is d^{10}. Thus *d–d* transitions are not possible for Cu^+.
This is an example of **charge transfer**.

Mn(VII) and Re(VII) have d^0 electronic configurations so these cannot be *d–d* transitions.

Mn(VII) is more easily reduced than Re(VII) (Mn(VII) is a stronger oxidizing agent) so electrons are more readily transferred from the O atom. Thus the transition for MnO_4^- is at lower energy or longer wavelength and occurs in the visible region. This is another example of charge transfer.

Ag^+ has a d^{10} electronic configuration so these cannot be *d–d* transitions cf. Cu^+ in part (b).

Electrons are more readily transferred from the more polarizable I atom so the transition is in the visible. A similar argument explains the colour of SnI_4.
$[Mn(H_2O)_6]^{2+}$ is octahedral and has a d^5 high spin electronic configuration. So *d–d* transitions are forbidden on all of Laporte, parity, and spin selection rules. $[MnCl_4]^-$ is d^4 and tetrahedral. As such the *d–d* transition is allowed by both the parity and spin selection rules.

Question S2.3

(a) Paramagnetic; d^9—odd number of electrons.
(b) Paramagnetic; odd number of valence electrons (11).
(c) Paramagnetic; even number of electrons (d^8) but tetrahedral structure, therefore two unpaired electrons.
(d) Diamagnetic; even number of electrons (d^8) but square planar and all electrons are paired
(e) Diamagnetic; as in part (d) square planar d^8 complex.
(f) Paramagnetic; even number of electrons but two unpaired electrons occupy degenerate pair of π^* orbitals.
(g) Paramagnetic; even number of electrons but two unpaired electrons occupy degenerate pair of π orbitals—note that σ–π crossover occurs in this example.
(h) Diamagnetic; even number of electrons. HOMO is $\sigma(2p_z)$ orbital because σ–π crossover occurs. This orbital contains a pair of electrons

CHAPTER 3

Question S3.2

(a) Higher nuclear charge.
(b) Pb^{2+} has largely ionic chemistry; Pb^{2+} is a relatively large ion (119 pm) and can accommodate nine Cl^- ions; Pb(IV) is much more polarizing and shows covalent chemistry.
(c) VSEPR shows that $PbCl_4$ has no lone pairs.
(d) In this instance VSEPR shows that $PbCl_2$ has a lone pair.

CHAPTER 4

Question S4.1

(a) i. Increased nuclear charge, not compensated by shielding.
ii. The outermost electron is further from the nucleus.
iii. The effect of Pairing and Exchange Energies.
(b) Although the first electron affinity of Na is exothermic the value is small. To form an ionic compound of Na^- would require the formation of a cation and this would be an **endothermic** process which cannot be compensated for by lattice enthalpy.

(c) In this case although it requires energy to form the gas phase anion O^{2-} this energy can be overcome by the lattice energy of the ionic product.

Question S4.2

(a) $MSO_4\,(s) \rightarrow MO\,(s) + SO_3\,(g)$

(b) This should show that CaO has the highest lattice enthalpy and $BaSO_4$ the lowest. The order should be CaO > BaO > $CaSO_4$ > $BaSO_4$.

(c) There is a bigger change in lattice enthalpy on going from $CaSO_4$ to CaO so the reaction is more favourable.

CHAPTER 5

Question S5.1

(a) Pyramidal structure. S has six valence electrons and is using four in bonding.

(b) Hard Lewis acids (smaller, higher oxidation state) tend to react faster and form stronger bonds with hard Lewis bases (less polarizable, higher electronegativity). Likewise soft Lewis acids (larger, lower oxidation state) tend to react faster and form stronger bonds with soft Lewis bases (highly polarizable and low electronegativity).

(c) The S atom is more likely to coordinate to Cd^{2+} and the O atom to Fe^{3+}.

(d) The central C atom uses all its electrons in bonding so it will be a linear ion.

(e) Fe^{3+} is a hard acid so the N atom will coordinate preferentially.

Question S5.2

a)

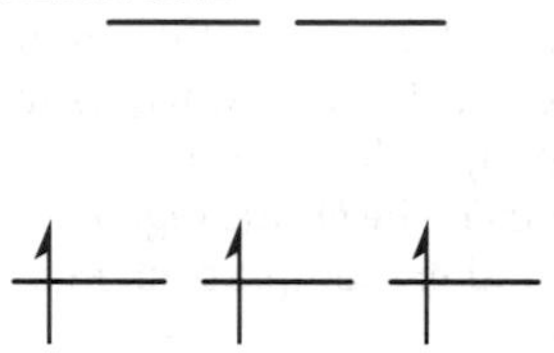

d^3 octahedral

b) NH_3 is a stronger field ligand than H_2O so Δ_o increases. Absorption shifts to higher frequencies (shorter wavelengths) and so the light transmitted will be of lower energy and longer wavelengths and so will change from green to purple.

c) $[CrO_4]^{2-}$ ion is d^0 and so the colour cannot be due to a *d*–*d* transition, therefore charge transfer likely.

d) An equilibrium is set up:
$2\,CrO_4^{\ 2-} + 2\,H^+ \rightleftharpoons Cr_2O_7^{\ -} + H_2O$

e) Zn powder reduces chromium(VI) to Cr(III) which is green.
$Cr_2O^{2-}_{7(aq)} + 14\,H^+_{(aq)} + 3\,Zn_{(s)} \rightarrow$
$2\,Cr^{3+}_{(aq)} + 7\,H_2O_{(l)} + 3\,Zn^{2+}_{(aq)}$

Index

Page numbers in bold type refer to sidenotes.
Page numbers italics refer to figures.

I

J

K

L

M

N

O

P

Q

R

S

T

U

V

W

X

Z